MAD Cats

The Story of Patrol Squadron 63
(VP-63)

By
Kernan Chaisson

DEFENSE
LION
PUBLICATIONS

MAD Cats

Acknowledgements

MAD Cats, the story of Patrol Squadron 63 (VP-63) is based on the reminiscences and other contributions of veterans from that unit. In their own words, they tell the story of this ground-breaking squadron and provide a unique and irreplaceable insight into a world that has now passed into history. It is truly an honor to help preserve the stories told by these veterans for future generations.

To Nadine, who makes everything worthwhile.

Copyright Notice

Copyright © 2012 Lion Publications Inc, 22 Commerce Road, Newtown, Connecticut 06470. ISBN 978-0-9859730-7-0 No part of this compilation may be reproduced or transmitted in any form or by any means, electronic or mechanical photocopying, recording or by any information and retrieval system without permission in writing from the publisher

The Story of VP-63

Contents

Preface	v
Introduction	1

Part One

Chapter 1. The Problem Develops	15
Chapter 2. Looking for a Solution	25
Chapter 3. A Stumbling Start	29
Chapter 4. A Good Plan Comes Together	39
Chapter 5. The Cats Draw Blood	47

Part Two

The MAD Cats – The Story of Patrol Squadron 63	53
Tall Tales	77
A Little Bit Of History	197
Return To RAF Pembroke Dock	201
Stained Glass Window	203
Flight Log Sniffer Wagner Spears, Mascot 1c VP-63	209
Appendices	221
A. Chronology	223
B. History Letter	225
C. Navy Press Release	235
D. Ode to a PBY	237
E. PBY-5/5A Specs.	239
Bibliography	241

MAD Cats

The Story of VP-63

Preface

War has always had a major impact on the world. Besides the basic result of breaking things and killing people, war has given birth to many world-changing technologies. The club, longbow, gunpowder; all were the result of seeking more effective ways of doing the basic business of combat. The march of time saw the growth of technology, and with it new and marvelous things. World War II was a watershed for the birth of many new tools of war. Radar was built on the fledgling science of radio and airplanes went from wood-and-wire oddities to powerful aluminum combat tools.

It is not so much what the sailors of VP-63 did, but what they started. We look at this through the eyes and lives of the naval aviators who gave birth to a major technology that came out of the war; something that became a mainstay of the anti-submarine effort to this day. A Magnetic Anomaly Detector (MAD), sometimes referred to as the Magnetic Airborne Detector, is carried by every U.S. and most major nations' ASW fleet, the 'stinger' sticking out the back of every P-3 and P-8A.

This is the way it is today. How we got there is the subject of this story. Source: U.S. Navy

MAD Cats

The German Navy *Kreigsmarine* was using the *unterseeboot* to choke off critical supply life lines between the United States and Great Britain. Stealth, the ability to attack and evade while submerged, made it possible for the 'under sea boats' to sink hundreds of ships, destroy thousands of tons of vital supplies, and take countless lives. The overall situation was grim. Old ways, limited detection capability, ineffective weapons; nothing seemed to overcome the German advantage. Something had to be done. Command dissention about how to solve the problem added confusion and let the carnage continue. Dependence on convoy tactics was not the whole answer.

When U-boats started operating out of French ports, the threat they posed grew a whole lot worse. Source: Deutsche Bundesarchive

A decision to give the problem to a small group of engineers and scientists was a giant step forward. Once an approach was born, proving the solution fell to a Navy patrol squadron of about 200 officers and sailors pulled from an assignment to the Pacific theater. With their PBY-5 flying boats, VP-63 crossed the Atlantic to prove that the new invention worked and, when combined with retro-bombs, could be used to sink submarines.

The Story of VP-63

As is often the case, it was not until many years after a group's exploits that the cover was drawn back and their trials and successes revealed. In this case, squadron reunions brought the squadron together in San Diego, California, and Pensacola, Florida; where they could talk publically and planners had the foresight to capture their memories and pull them together in writing and on video. It saw the publication of yearbooks prepared for the reunion and a videotape of oral histories recorded.

Mostly young men who had volunteered to fight the war, they took on the job and its risks with enthusiasm and dedication, creativity and humor. It is important to make a distinction. Some organizations are important because of what they accomplished; others are important for what they started. VP-63's true value is more the latter than former. What they did during the war was noteworthy. It was not the number of submarines they destroyed, but instead because they showed how to fight them in a new way.

During an interview at a squadron reunion in San Diego in September, 1983, Andrew H. Reid, an original member of VP-63 spoke of a command he had following the war. "In 1957, I commanded a carrier squadron which had the identical equipment on board, used the identical tactics. . . . The squadron would not believe the equipment and tactics were created in 1942 because VP-63 was so very hush-hush. No one knew about us then and very few know about us now, even though everything has been de-classified. VP-63 deserves a little more than deep secrecy, but that was our reward."

In going through interviews, histories, and writings from the participants, it is obvious that at the time squadron members felt the importance of what they were doing. They also knew that much of what they did would remain hidden for a long time. Classification rules would keep their story from the public eye for decades and prevented them from sharing their experiences with friends and relatives, in all but the most rudimentary way. It was not until time caught up with classification that the men of VP-63 could begin to reveal the details of their stories.

MAD Cats

There is no shortage of written material about the fight against Germany's submarine force in WWII; but none are quite as personal as this one told from the standpoint of the officers, enlisted men, and one dog in a unique Naval flying squadron. VP-63 was a unit where many little non-glamorous acts led to big success. There was special value to letting each man tell his own version of parts of the story. The result was not just a history of one part of the war. Instead we get to live the times through their eyes.

This collection of stories from the members of VP-63, who lived and slept with the "MAD Cat" through almost three years of World War II, is not a gory tale, for VP-63 personnel normally did not see that kind of war, the agony of injury and death on the ground and in trenches. Theirs was a relatively clean war usually over the beautiful sea. When tragedy struck, the consequences were instant and final. So if the reader is expecting a blood and guts war story, this is not one. Many stories are humorous, with tragedy mostly in the background. In the main, the funny things that happened "along the way" and the special lives they forged there is what it is all about.

A special thanks goes to the men of VP-63 and their families. One was this writer's uncle, Adam Chaisson, radio and MAD operator, who entrusted me with copies of books prepared for the unit's 42nd and 43rd anniversary reunions as well as other material valuable in preparing this book. The classified nature of their mission made the official records available at the Naval Historical Center in Washington, DC, a bland collection of numbers and dates.

World War II unit reunions have always been popular, especially given the pressure of the calendar. They are a chance to renew connections, check the status of old buddies, find out who made the Final Change of Station. War stories abound as members re-live those old times. Fortunately for us, planners had the foresight to capture these memories in writing for us and for the future, and Lt. William P. Craddock took on the job of compiling them into *The History and Tales of the MAD Cats*.

The Story of VP-63

While there has always been an interest in World War II history, the calendar is generating pressure to capture the personal stories of the men and women who did what was needed, often with an amazing spirit and a unique sense of humor.

Combat can bring out the worst and best in those who fight it. While there are volumes studying the leaders and their tactics, the better stories can be found down at the lower levels where men and women fight the wars. During a writers' week program at The Chautauqua Institution in New York State in June, 2012, authors said that stories should create in peoples' minds images of the experience of others, letting us feel that experience. With that in mind, the bulk of this book capitalizes on what was written by the sailors of a long-unheralded group who reflect what is true and good in the American GI. This reflects on all of us, and hopefully speaks to us.

VP-63's wartime experiences, and therefore their stories, are something many of us can relate to. Some of their personal stories will no doubt touch deep-seated feelings and memories of personal experiences. While so much of what was written about World War II gives us hard facts and cold statistics, these stories instead are filled with feelings and ideals; they are from the heart.

The submarine fight of World War II was widespread and complicated. It included Pacific combat as well. This book makes no attempt to cover anything but one small, but important, part in Europe.

This accounting combines the technical and operational story in the personal words of the sailors and officers of the MAD Cats. It gives life to the joys, fears, victories, and losses that make up their tale. It takes advantage of personal and squadron logs. One especially interesting source was the flight log of the squadron's mascot, a mixed-breed dog named Sniffer Wagner Spears. He flew a reported 500 hours the first year, and was on board for one submarine sinking; carefully logging each flight. He also managed to get called before a Court Martial at one point – but more about that later.

MAD Cats

Acknowledging wartime exploits is often delayed, and the accomplishments of many never get the recognition they deserve. Like the technical 'Boffins' of England, VP-63's exploits were nearly missed by the searchlight of history. This is a chance to change that. While most writers focused on the operational history and tactical planning of the submarine war, Montgomery C. Meigs, in *Slide Rules and Submarines*, detailed the technical developments underlying the battles; as well as implications thereof. It fell to the scientists, engineers, and operators in organizations such as VP-63 to make countering the undersea threat possible.

This is their story – in their own words.

I would like to acknowledge the help of Keith Chaisson, who in the days before it became easy, scanned and converted the stories and texts into computer files that were used for much of the book; a massive effort. Mark T. Weber, Curator at the U.S. Navy Memorial Foundation, provided invaluable assistance by maintaining the original material that was donated by Adam J. Chaisson and filed at the Memorial's Reference Center in Washington, DC. Original copies of the squadron books, reunion programs, and a reunion video are available to the public in the Resource Center at the Navy Memorial on Pennsylvania Avenue in Washington.

As World War II becomes more distant, veterans of that war are disappearing. In 2012, the Veterans Administration estimated that 1.7 million were still alive; but they were passing away at the rate of nearly 250,000 a year; with the rate increasing as the vets age. The concern that the world would lose the stories of those who lived the experience, a national treasure, prompted Congress to establish *The Veterans History Project* in 2000. While The Library of Congress has an active, ongoing program to capture the stories of individual veterans while it can, it is feared that only one in two hundred stories will be preserved. Other, private efforts are encouraged.

This is one.

The Story of VP-63

Introduction

There can be no talk of a let-up in the U-boat war. The Atlantic is my first line of defense in the West. And even if I have to fight a defensive battle there, that is preferable to waiting to defend myself on the coast of Europe. The enemy forces tied down by our U-boats are tremendous, even though the losses inflicted by us are no longer great. I cannot afford to release these forces by discontinuing the U-boat war.

Adolf Hitler, May 21st 1943

In *Memoirs: Ten Years and Twenty Days* Grand Admiral Karl Doenitz, the commander of all Germany's submarine forces, wrote that during a conference with Adolph Hitler on 17 September, 1941, the *Fuehrer* gave instructions that all incidents with the United States were to be avoided. But the Admiral felt that Germany should take advantage of the element of surprise and strike in waters where the element of surprise was a positive factor.

Doenitz explained that the Japanese attack on U.S. forces on Pearl harbor December 7th, 1941, surprised the German Naval High Command and there was not a single U-boat in American waters. On 9 December, High Command lifted all restrictions and the Admiral requested twelve U-boats for operations off the American coast.

Merchant vessels sailed independently for Canada to such ports as Halifax and Sydney, Nova Scotia, where convoys heading for England formed. They were sitting ducks. Doenitz felt that conditions would be "at least as favorable for the conduct of U-boat operations as those…a year or two earlier in British waters. …. Although the British and Americans had been exchanging information to create a system of anti-submarine defense, due to a lack of practical experience it would not be very efficient. …. Sooner or later, of course, those favorable conditions would gradually disappear. …. It was, therefore, of primary importance to take full advantage of the favorable situation as quickly as possible and with all available forces, before the anticipated changes occurred." (*Memoirs*, p.196)

MAD Cats

In the Twin Towers Entertainment documentary video *U-boats: The Wolf Pack*, German U-boat Commander Captain Herbert Werner said that "The Second World War started at practically the same level as World War One in 1914. Countermeasures against U-boats were practically nil." British shipping was the first prey, then the Germans came to the East Coast of America.

In an interview, Captain John M. Walters, USCG (retired) explained that although "they never had more than 12 subs on station, 600 men, in six months (the Germans) sank 2 ½ million tons of shipping right along our coast. You could see the ships burning from the beaches."

Captain Waters went on to explain that as far as the submarine threat went, "we were simply unprepared for it. The Navy's eyes were on the Pacific and on the Japanese Fleet. It was two years of attacks on the East Coast before we saw what was happening in European waters. Simply, we were not prepared for war on our own doorsteps."

In the same video, Captain Herbert Werner said that "In the winter of 1942, German submarines sailed from concrete bunkers along the north coast of France, refueled in the Atlantic by 'milk cows', and were controlled by short-wave and ultra-long-wave radio. There were no defenses for convoys off Labrador and Greenland." Captain Werner went on to say that "once, we reached the east coast of the U.S. there were no countermeasures. We even penetrated the Chesapeake without seeing one US Navy vessel."

The German submarine threat along the Eastern Seaboard and in the Caribbean eventually generated desperate efforts at countering the menace. Admiral Doenitz and the 'Happy Times', wreaked havoc on Allied shipping, giving rise to all manner of attempts to counter the submarine threat. While many solutions panned out, there were others that were not all that great.

During a combined Smithsonian Associates/International Spy Museum course on WWII spies, Dr. Nicholas Reynolds, Historian at the CIA Museum, related the story of famous author Ernest Hemingway's personal plan for countering the menace. For a while, he circulated around Cuba to gather intelligence on Germany,

The Story of VP-63

passing what he found to the American embassy there. A special unit was formed to handle this intel, a group Hemingway called "The Crook Factory". The idea may have been good, the results not so much.

Hemingway, who tended to be mostly interested in fishing and drinking – sometimes with a Tommy Gun (honest, there are photos) – provided reports that were something less than stellar. Among the information passed on by the embassy was that there were 1,000 German submarines operating in U.S. waters. These false and outrageous reports created confusion and false leads. Naval officials had enough to deal with without this muddying the waters. In his memoirs, Admiral Doenitz wrote that due to the long voyage from the German submarine pens there were never more than ten or twelve boats in American waters at one time.

In 1942 and 1943, Hemingway volunteered to sail the waters north of Cuba to look for subs and sink them. His fishing boat *Pilar* would look as if it were selling water and fish. According to Hemingway's plan, when a sub pulled along side, jai lai players he hired would toss hand grenades down the hatch and machine gun the crew. He may have sighted one submarine total, according to Dr. Reynolds.

Fortunately, saner heads prevailed and other approaches came to the fore; including MAD and other attempts at technical solutions to the anti-submarine problem.

Interestingly, even though these events took place over six decades ago, interest in continued unabated. In July, 2012, the wreck of U-550 was discovered off the coast of Massachusetts about 70 miles south of Nantucket. A team of divers, lead by a New Jersey attorney located the wreck after a multi-year search.

ASW Develops

Anti-submarine warfare (ASW) is a branch of naval warfare that uses surface warships, aircraft, or other submarines to find, track, and deter, damage or destroy enemy submarines. According to history books, the first attacks on a ship by an underwater weapon are generally

MAD Cats

believed to have been during the American Revolutionary War, using what would now be called a naval mine but what then was called a torpedo. The first self-propelled torpedo was invented in 1863 and launched from surface craft. The first submarine with a torpedo was *Nordenfeld II* built in 1886, though it had been proposed earlier. In the Russo-Japanese War of 1904-5, the submarine became a threat. By the start of the First World War nearly 300 submarines were in service. Some warships were fitted with an armored belt as protection against torpedoes.

Montgomery Meigs, in the National Defense University book *Slide Rules Versus Submarines*, explained how the Allies prepared for, or in some cases failed to prepare for the trials of a German submarine menace. Like many forms of combat, successful anti-submarine warfare depends on a mix of sensor and weapons technology, training, experience, and luck. Sophisticated sonar equipment for detecting, classifying, locating, and tracking the target submarine is a key element of ASW. To destroy submarines, both the torpedo and mine are used; launched from air, surface and underwater platforms. Other means of destruction have been used in the past but are now obsolete. ASW also involves protecting friendly ships.

U-boats created havoc when used against British supply lines in 1917. Source: Deutsches Bunderarchive

In the beginning, there were no means to detect submerged U-boats, and attacks on them were limited at first to efforts to damage their periscopes with hammers. The Royal Navy torpedo establishment, HMS *Vernon*, studied explosive grapnel sweeps; these sank four or five U-boats in the First World War. A similar approach featured a string of 70 lb (32 kg) charges on a floating cable, fired electrically. An unimpressed Louis Mountbatten felt that any U-boat sunk by it deserved to be.

The Story of VP-63

During the First World War, submarines were a major menace. They operated in the Baltic, North Sea, Black Sea, and Mediterranean as well as the North Atlantic. Previously they had been limited to relatively calm and protected waters.

One approach to countering submarines was to use a range of small, fast surface ships using guns and luck. They mainly relied on the fact a submarine of the day was often on the surface for a range of reasons, such as charging batteries or crossing long distances. The first approach to protect warships was chain link nets strung from the sides of battleships as defense against torpedoes. Nets were also deployed across the mouth of a harbor or naval base to stop submarines entering or to stop torpedoes fired against ships. British warships were fitted with a ram with which to sink submarines, and U-15 was sunk in August 1914.

Seaplanes and airships were also used to patrol for submarines. A number of successful attacks were made; but the main value of air patrols was in driving the U-boat to submerge, rendering it virtually blind and immobile. However, the most effective anti-submarine measure was the introduction of escorted convoys, which reduced the loss of ships entering the German's War Zone around the British Isles from 25% to less than 1%. (*Slide Rules*, p 6)

Meigs wrote that according to records in the National Archives and Research Agency, Record Group 227, the Records of Division 6 of the National Defense Research Committee, by the end of the WWI, the toll on U-boats was 30% lost to mines on North Sea ingress and egress routes, 10% sunk by Allied submarines on the surface, 25% rammed by surface ships or destroyed by gunfire, 20% sunk while submerged, and 15% were lost to other Allied action. (*Slide Rules and Submarines*, p 6).

To attack submerged boats, a number of anti-submarine weapons were derived; including the sweep with a contact-fused explosive. Bombs were dropped by aircraft and depth charge attacks were made by ships. Initially, these were simply dropped off the back of a ship; but then depth charge throwers were introduced. The Q-ship, a warship disguised as a merchantman, was used to attack surfaced U-boats

MAD Cats

while the R1 was the first ASW submarine. A major contribution was the interception of German submarine radio signals and breaking of their code by "Room 40" of the Admiralty.

603 of 863 operational U-boats were sunk during the war from a variety of ASW methods. (*Memoirs*, p.489)

The Magnetic Anomoly Detector

A Magnetic Anomaly Detector (MAD) is an instrument used to detect minute variations in the Earth's magnetic field. The term refers specifically to magnetometers used by military forces to detect submarines (a mass of ferromagnetic material creates a detectable disturbance in the magnetic field). The military MAD gear is a descendent of geomagnetic survey instruments used to search for minerals by the disturbance of the normal earth-field. Geoexploration as a result of measuring and studying variations in the Earth's magnetic field has been conducted by scientists since 1843. The first uses of magnetometers were for the location of ore deposits. Thalen's *The Examination of Iron Ore Deposits by Magnetic Measurements*, published in 1879, was the first scientific treatise describing this practical use.

In *The U-boat Hunters,* Anthony J. Watts wrote that "during the winter of 1939/40 a number of experiments were conducted at Farnborough (the British defense research facility near the town of Bagshot) which showed that the changes in the earth's (magnetic) field caused by the passage of a submarine through the water were so small that with the detection apparatus then available the idea was impractical for the British to develop. The idea was taken up by the USA." (*U-Boat Hunters,* p.149)

There was some misunderstanding of the mechanism of detection of submarines in water using the MAD boom system. Magnetic moment displacement is ostensibly the main disturbance; yet submarines are detectable even when oriented parallel to the Earth's magnetic field, despite construction with non-ferromagnetic hulls. For example, the Soviet-Russian *Alfa* class submarine, whose hull is constructed out of titanium to give dramatic submerged performance and protection from

The Story of VP-63

detection by MAD sensors, is still detectable. This is due in part to the fact that even submarines with a titanium hull will have a substantial content of ferromagnetic materials. The nuclear reactor, steam turbines, auxiliary diesel engines, and numerous other systems will be manufactured from steel and nickel alloys.

In early 1941, the U.S. National Defense Research Committee assumed sponsorship of the device and had demonstrated the ability to locate a submarine from an aircraft 400 feet above. Because it was a passive technique, the U-Boat crew would not realize they were under surveillance or had been detected until attacked.

By autumn of 1942, U.S. scientists had advanced their efforts on the MAD device. Before the war, mineralologists had used magnetometers to plot distortions in the earth's magnetic field and locate underground mineral deposits. Although these devices were sensitive enough for mineral exploration, they could not detect a small object like a submarine. In 1940, Victor Vacquier of the Gulf Research and Development Company in America (which became the Gulf Oil Corporation) produced a 'saturable-core' type of magnetometer (US Patent 2,406,870). (*Aircraft Versus Submarine*, p 99) It was two or three times as sensitive as the prospecting version. Because unlike radio waves a magnetic field is not effected by passing through the interface between air and the sea, engineers realized its potential for detecting submarines.

The strength of the earth's field varies with latitude, but a typical figure is 50,000 gammas. The magnetic field of a typical 700 ton WWII submarine was in the order of ten gammas at a distance of 400 feet. This field strength decreases with the cube root of distance, so at 800 feet the field reduced to 1.25 gammas. Moreover, the magnetic detector had to be aligned with the earth's magnetic field to within one tenth of a degree, or its sensitivity fell drastically. Yet another problem was interference from the metals in the airplane itself. This was solved by placing the detector in a boom extending from the tail and replacing ferrous components in the vicinity of the detector with non-ferrous materials. Because of these restrictions, a searching aircraft had to fly directly overhead at an altitude of 100 feet or less if

MAD Cats

it was to detect a submarine at a depth of 300 feet. The result was the inability of using the device at night or in bad weather. (*Aircraft versus Submarines*, p 99)

Interestingly, what goes around tends to come around. At the MAD Cats reunion in an article was included in the reunion book citing *The Houston Post* Energy Writer Sam Fletcher discussing how serchers were now using micromagnetics to search for petroleum. He cited how "a magnetometer – the most sensitive ever invented – mounted in the 'stinger' tail of a twin-engine airplane is flown at speeds of about 130 knots 200 feet above the terrain, recording data in quarter-mile swaths."

"VP-63 Lives On," a squadron member wrote in the margins of the article.

Early in 1942, two American firms, the Western Electric Company and the Airborne Instruments Laboratory, began work on producing this submarine detection equipment for aircraft. During the spring the first trial sets took to the air in Naval patrol aircraft and airships, and later in the year the device entered service in units patrolling the waters along the Eastern Seaboard of America. Initially the device could do little, because by that time the German submarines had shifted their attack elsewhere. Improved convoy tactics and a more aggressive use of scout aircraft and surface ships were making life difficult for the *unterseeboots*.

The operational employment of MAD introduced problems of its own: the range of the equipment was so short that the operator received an indication of the presence of a submarine only when he was almost immediately overhead. If he released ordinary depth charges based on the magnetic detector information, their forward speed as they left the airplane would carry them over and clear of the submarine. What was needed to complement the MAD was a special bomb, which would be released at an aircraft speed of 100 mph but still fall vertically. To meet this requirement the California Institute of Technology produced the so-called 'retrobomb', a 35 pound impact-fused explosive device, with a solid fuel rocket fitted into the tail. When the MAD equipment indicated a submarine

The Story of VP-63

below, the aircraft operator pressed a button to fire the retrobombs. The rockets propelled the bombs backwards off their launching rails at about 100 mph. This brought them to a stop in mid-air. The rockets, their work done, ceased burning and the bombs fell vertically into the sea. (*The U-boat Hunters*, p 150-151)

A PBY-5 *Catalina* could carry twenty-four of the small retrobombs, twelve under each wing on special launch rails. When the operator pressed the button to fire the bombs, the first salvo of eight roared off their rails; half a second later the second salvo of eight followed automatically, and half a second after that the third salvo went.

Retrobombs under a MAD Cat's wing. Source: U.S. Navy

The retrobomb launchers were divided into eight groups of three, each group being set at a slightly different angle; as a result the bombs from each salvo hit the water in a line about 180 feet long, perpendicular to the line of flight of the aircraft. The salvoes spread about 100 feet, with the half-second delay between salvoes giving a spacing of about 90 feet between these lines. Prior to attacking the submarine with bombs, the aircraft crew tracked its movement with smoke markers; as a result they would be able to attack the boat along its length, and thus stand a good chance of scoring a potentially lethal blow with at least two of the retrobombs.

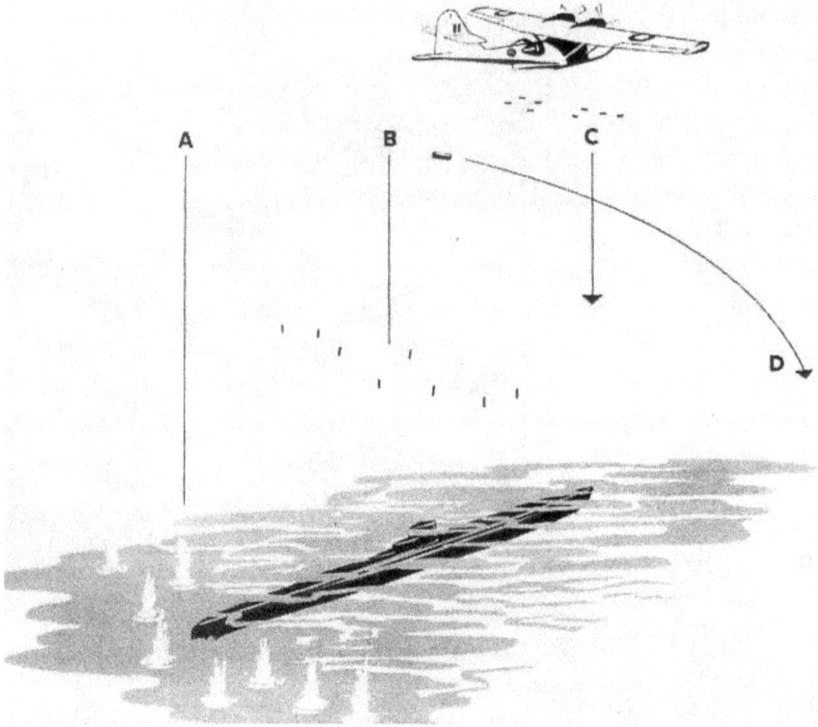

MAD Cats

Retrobombing tactics. Source: U-boat Hunters, p 103

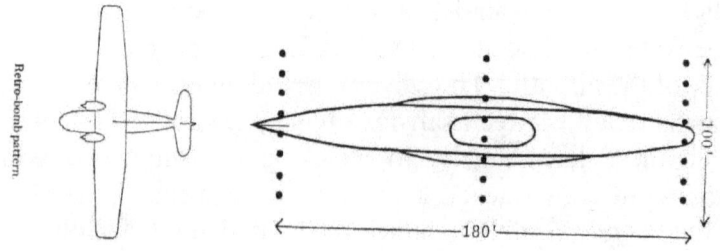

Retrobomb Release/Strike Pattern: Source: Aircraft versus Submarines, p105

The Story of VP-63

Of course, with the MAD gear at the time it was very difficult to tell whether a metallic object found was in fact a submarine; it could, for example, be a far larger wreck lying on the seabed. The need to resolve this problem hastened the development of another new device, the sonobuoy. The sonobuoy consisted of a small floating radio transmitter, under which a hydrophone was suspended on a length of line. Listening for any sounds in the water, the hydrophone passed them to a transmitter on the surface, and the aircraft picked up the buoy's transmissions with a special receiver.

Although the German Navy suspected the existence of such equipment, there was no evidence of its use by the Allies; the suspicions were aired at a conference held on 10th March 1944 when the head of the development section of the German Naval Technical Signals Department, Kapitan zur See Helmuth Giessler, informed the assembled officers: "While we employ only acoustic location devices [to find submerged submarines] there is a suspicion, which we are examining carefully, that perhaps the enemy is using magnetic-field-change methods as well. The range of such magnetic locators, from ships or from aircraft, is now being examined; from aircraft, a range of between 200 and 400 metres would appear to be theoretically attainable.

"Giessler's statement illustrates, yet again, the paucity of German Intelligence on the Allied anti-submarine measures: by this time the U.S. Navy had had such a magnetic device in service for over a year, and only a fortnight earlier it had played an important part in the destruction of a U-boat". (*Aircraft versus Submarine,* p. 188)

Statements from the U-boat commander of one of the boats sunk by the MAD Cats revealed complete surprise at the attack. He seems to have had no clue as to what found him.

In his *Memoirs*, Admiral Doenitz only mentioned radar, direction-finding, and Leigh Lights as keys to the ability of Allied forces to find submarines. He admitted that improvements in locating devices and increased air cover filled him with "grave misgivings." Not a

MAD Cats

word about MAD. In the Afterword, written with the advantage of history, MAD is not mentioned either.

U-boat Commander's Handbook

Interestingly, nowhere in the 1943 version of the German *Submarine Commander's Handbook*, translated from a copy found in U-505 that was captured intact on 4 June, 1944, is there any discussion of how to avoid detection by MAD gear. "Much more than a guide, it was a bible, giving exact counsel for every situation anticipated by the High Command of the Navy. The handbook provided detailed guidance for every situation, every maneuver," wrote E.J. Coats in the January, 1989, edition published in English.

VP-63 PBY-5A Catalina.
Source: U.S. Navy

The handbook states that key to a submarine's success is the ability to attack without being detected. In order to remain undetected, before and during the attack the submarine must neither be sighted nor sound-located, nor detected by ASDIC (sonar)," the Handbook counseled. The submarine captain's guide called aircraft the U-boat's greatest enemy, visual sighting or radar were considered the main threats. Subs were advised, especially during operations where it had to remain stationary, to submerge from dawn to dusk. It should also pass through danger areas at periscope depth – both actions making detection by MAD more likely. Periscope use should be planned to prevent what was considered the most likely way an aircraft could find a U-Boat, visual detection.

One paragraph notes that when an airplane is almost vertically above a submerged submarine, because of the high speed of the aircraft is was very difficult to spot a moving U-Boat. This overhead position is the sweet spot for MAD detection. One captured U-Boat captain could

The Story of VP-63

not understand how VP-63 had managed to find and attack his submarine. ASDIC was feared, but it could be detected by the submariners and his crew had heard nothing.

Radio-Direction-finding (RDF) was also blamed for the detection of U-Boats. But through the end of the war it did not seem that the German undersea fleet ever realized that Allied MAD existed, much less posed a threat. Some German engineers doubted that such a detection scheme was even possible.

The Catalina's phenomenal range made it a natural platform for ASW operations. However, the amphibious PBY-5A version shown here suffered due to the extra weight of its land undercarriage and most MAD Cat crews preferred the non-amphibious PBY-5.
Source: U.S. Navy.

There was one significant mission performed by a *Catalina* during the "Bloody Winter" of 1942-43. In late December, a convoy departed Iceland to join up with its submarine protection ships at the ICOMP (Iceland Convoy Oceam Meeting Point.), where they would join and head for Europe. Bad weather and other forces prevented the ships from meeting up as they were supposed to. They could not send or

MAD Cats

receive radio messages to coordinate because of the dire threat of the Germans direction-locating the ships and attacking. If nothing could be done, however, radio silence would have to be broken, to the detriment of all ships involved.

On December 14th, the weather cleared enough for a PBY to be sent from Iceland. Although this would be at the extreme range of the aircraft, there was hope that it could help. When the aircraft was sighted from the convoy and identified as a PBY, the Skipper of the escort ships was reported to say, "Good old VP-73 (VP-63's sister squadron). This guy's a long way from home. Circling low, the PBY established blinker light contact with the ships and was able to report the location of the convoy, making a rendevous possible. (*Bloody Winter*, p 95-96)

The Story of VP-63

CHAPTER 1

The Problem Develops

World War I changed and shaped modern warfare; but mainly from the standpoint of land campaigns. The machine gun, modern artillery, the airplane, the tank, and the telephone changed the face of war. At sea, battleships reigned. The effective aircraft carrier did not begin to develop until well after the war. Germany mounted a submarine campaign in 1914; but the U-boats operated as surface raiders. They submerged to sneak up on prey. When they intercepted merchant vessels, the submarine would surface and issue a warning, giving the crews a chance to abandon ship; after which they sank the vessel with their deck gun. Torpedoes were saved for the *coup de grace*.

U-boats started out as fleet scouts but found their true calling as commerce raiders. Source: U.S. Navy

Surface ships used guns or ramming to attack submarines on the surface at first, in keeping with the international Prize Ordinance agreement. But the German U-boat commanders began to stray from the rules of maritime warfare and began conducting submerged attacks. In 1917, the German High Command initiated unrestricted submarine warfare. The U-boats began to attack while submerged and without warning. In April 1917, the British Merchant Marine lost 900,000 tons, one twentieth of its capacity, to U-boats. Allies reverted to convoys and escorts for protection.

There was no particular technology dominant in the anti-submarine warfare arena. Well trained and aggressive escorts kept the subs away

MAD Cats

from convoys and an occasional lone raider was discovered. Airplanes were considered as possible ASW platforms; but the war ended before this concept had been explored very much. Submarine detection tended to be based on direction finding of radio communications as well as familiarity with operational techniques which allowed command centers to predict U-boat movement so convoys could be routed away from where contact was likely. Forcing submarines to submerge to avoid detection lessened their effectiveness against a convoy.

The British had developed an underwater sound ranging system called ASDIC, while the United States had begun work on hydrophones for capturing underwater sounds. Crude, second-generation versions of this equipment had been fielded by the end of WWI. Improvements in depth charges and better sub chasers and submarines were found; but once the war ended, interest waned and progress ended.

In WW1, the Felixstowe F.2A proved effective at keeping submarines submerged but lacked weapons and sensors for attack. Source: U.S. Navy

No agency survived World War I with the power and interest necessary to carry on effective anti-submarine warfare developments. Most efforts produced incremental improvements to existing equipment, but did not specifically focus on new operational needs or concepts. For example, the Naval Research Laboratory perfected underwater sound ranging gear; but there was a limited effort to integrate this improved equipment into an better ability to find and kill submarines. Nor was there much effort to pull together a training program to make ASW work. Anti-submarine warfare weapons had actually improved only marginally over those available in 1918. (*Slide Rules* p 10). Combat plans focused on how

The Story of VP-63

to best employ surface fleets to achieve specific objectives. Airplanes were scouts for battle groups while submarines were considered ideal for reconnoitering for surface ships and only secondarily for attacking them.

The US strategy, even up until 1940, was based on the idea that the secret to naval victory was the fire power of the surface fleet. Control of the sea depended on the offensive action of battleships and the fast aircraft carrier groups. Submarines and aircraft supported these fleets; but they had no independent function.

Things were no different in Britain. Strategists confined themselves to combat between big ships. Although they had improved the depth charge in 1925, the British did not consider dropping them from aircraft. (*Slide Rules*, p 10). In fact, the British were said to have made an affirmative decision not to develop an anti-submarine bomb.

Germany had a different idea. Although losing WWI slowed their ability to build a large fleet, planners remembered that in 1917 submarines nearly severed Great Britain's maritime lines of supply. A loophole in the Anglo-German Naval Treaty put submarine development into a special category and opened the door for the modernization of the German undersea fleet. Toward the end of the war, a young U-boat captain, Karl Doenitz, successfully employed surface attacks at night and began to develop the idea that submarines operating in packs could wreak havoc on convoys and escorts. Doenitz was to become the father of the German U-boat navy in World War II.

Admiral Karl Doenitz.
Source: Deutsche Bundesarchive

In July, 1935, the German High Command appointed Karl Doenitz Officer Commanding Submarines. Since the end of World War I, Admiral Doenitz had been studying surface operations. Combining

MAD Cats

this knowledge with his undersea experience, he developed operational concepts that he hoped would prove effective against the Allies once war began. He took advantage of peacetime to train and test his U-boat crews. By 1937 he had developed the skills and confidence in his crews that made wolfpack tactics a success. The die was cast and Germany had created a problem unanticipated by the Allies and a solution what would eventually involve VP-63.

Meigs went on to explain how in the post-war period, budget restrictions limited the construction of new ships and naval planning focused on the traditional, clashes of surface fleets. (*Slide Rules*, p 6). As a result, the Allies entered World War II relatively unprepared for anti-submarine warfare. Neither the Americans or the British had developed a new operational concept for an anti-submarine campaign.

At the beginning of the war, most navies had few ideas how to combat submarines beyond locating them with sonar and then dropping depth charges. Sonar proved much less effective than expected, and was of no use at all against submarines operating on the surface, as U-boats routinely did at night. The Royal Navy had continued to develop indicator loops between the wars, but this was a passive form of harbor defense that depended on detecting the magnetic field of submarines with long lengths of cable laid on the floor of the harbor. Indicator loop technology was developed further and deployed by the US Navy in 1942. By then there were dozens of loop stations around the world, but Sonar was far more effective and the use of loop technology faded.

One bit of good fortune for the Allies, the war started in 1939, before Doenitz could develop the fleet he wanted. Instead of 300 ocean-going U-boats, he had 60 submarines, 30 of which could conduct prolonged operations at sea. Only ten were of the new 500-ton class.

In September, 1939, the Germans began their attacks on British shipping. They had developed a wakeless torpedo; but had not seriously worked on developing a scientific solution to keeping the boats hidden. They also did not anticipate or provide protection from the combined effects of radar, sonar, and the airplane. (*Slide Rules*, p. 17) But though smaller than he wanted it to be, Doenitz's U-boat

The Story of VP-63

force was well trained, confident, combat ready. Their target was the British sea lanes on which the nation's very existence depended.

In his book *The Double-Cross System in the War of 1939 to 1945*, J.C. Masterman noted that German questions to their in-country spies in Britain in 1940 focused on foodstuffs and commodities available to the population and the impact shortages would have on the morale of the populace. Nazi planners wanted to know how much was arriving from Canada, the United States, and South America to get a feel for the impact of their blockade. Questions asked of the espionage system can be a good indication of enemy intent and strategy. These showed the importance of an effective submarine campaign against England's supply lines for civilian goods as well as war materiel.

By the end of 1941, Germany had tripled the number of U-boats on patrol to 30. But the Royal Navy had improved its ability to protect convoys end-to-end from Halifax, Nova Scotia to England. From April to December 1941, U-boats sank an average of 34 merchant vessels monthly, with a loss of only 2 to 3 submarines. In December 1941, in heavy weather, escorts of convoy HG 76 soundly defeated the attacking U-boats. The success rate dropped to one merchant ship per month per submarine on patrol in the North Atlantic. But independent, unescorted shipping was a lucrative target. To achieve the goal of sinking as many merchant ships as possible at the smallest cost in submarines, in early 1942 Admiral Doenitz shifted the focus of the attack to the unprotected American East Coast.

The effect was terrible. The scientists had been working on a variety of devices to counter the submarines, countermeasures unknown to U-boat command. But none were ready to be put into the field. Doenitz was bolstered by his evaluation that the submarine's greatest asset, the element of surprise, remained. He felt that on the surface, U-boats could not be spotted quickly enough for targeted ships to avoid them. When on the attack, they could not be detected early enough by surface or underwater means to negate the attack.

On 12 January 1942, Doenitz ordered six Type IX U-boats into American waters where the crews pressed the attack. U-boats ranged down the great circle route into the shipping lanes of New York,

MAD Cats

Hatteras, and Florida. In February, the Allies lost 85 ships, 570,000 gross tons. 90% of them went down off the coast of North America. Doenitz's goal was 700,000 tons of losses on the Allies monthly.

Not everything went the U-boat leader's way. The German Admiral could not get the priority for submarine construction he wanted, slowing the rate at which he could expand his patrols. The German High Command also required that he keep submarine patrols off Gibraltar and Norway, diluting the number of U-boats available to attack allied shipping along the American East Coast. Finally, the Luftwaffe would not give him the aircraft needed to control the airspace over the Bay of Biscay. This would come to have a direct impact on VP-63.

Comfortable with his successes, Doenitz did not dig too deeply into the application of technological innovation to subsurface warfare. A year later in the Atlantic, the German U-boat Command would be faced with the results of the American scientists' work. By then it was too late to start fielding weapons and tactics to counter what the Americans, including VP-63, were throwing at them.

One of the best kept Allied secrets of the war was the breaking of enemy codes, including the German Naval Enigma codes (information gathered this way was dubbed *Ultra*) at Bletchley Park in England. This enabled tracking U-boat packs to allow convoy re-routings; whenever the Germans changed their codes (and when they added a fourth rotor to the Enigma machines in 1943), convoy losses rose significantly. By the end of the war, the Allies were regularly breaking and reading German naval codes.

To prevent the Germans from guessing that Enigma had been cracked, the British planted a false story about a special infrared camera being used to locate submarines. The British were subsequently delighted to learn that the Germans responded by developing a special paint for submarines that exactly duplicated the optical properties of seawater.

Many different aircraft, from airships to four-engine sea- and land-planes, were used in the submarine fight. Some of the more successful were the Lockheed Ventura, PBY (Catalina or Canso in British

The Story of VP-63

service), Consolidated B-24 Liberator (VLR Liberator, in British service), Short Sunderland and Vickers Wellington.

The Short Sunderland was an effective and popular maritime patrol aircraft. Source: Royal Air Force

U-boats were not defenseless, since their deck guns could be a good anti-aircraft weapon. They claimed 212 Allied aircraft shot down for the loss of 168 U-boats to air attack. From August 1842 to May 1943, German wolfpack tactics from the East Coast of the United States to the Bay of Biscay sank 3,857,705 tons of Allied shipping, at the cost of 123 U-boats. At one point in the war, there was even a 'shoot-back order' requiring U-boats to stay on the surface and fight back, in the absence of any other option.

It turned out that fighting it out on the surface was a really bad idea. Source: Deutsche Bundesarchive.

Providing air cover was essential. The Germans at the time had been using their Focke-Wulf Fw 200 *Condor* long range aircraft to attack shipping and provide reconnaissance for U-boats, and most of their sorties occurred outside the reach of existing allied land-based aircraft that the Allies had; this was dubbed the Mid-Atlantic gap. At first, the British tried temporary solutions such as CAM ships (CAM ships were British merchant ships used in convoys as an

MAD Cats

emergency stop-gap until sufficient escort carriers became available. CAM is an acronym for Catapult Aircraft Merchantman. They were replaced by mass-produced, relatively cheap escort carriers built by the United States and operated by the US and Royal Navies. There was also the introduction of long-range patrol aircraft. Many U-boats feared aircraft, as the mere presence would often force them to dive, disrupting their patrols and attack runs.

The CVE escort carriers proved to be a vital link in the protection of transatlantic convoys. Source: U.S. Navy

There was a significant difference in the tactics of the two navies. The literature shows that the Americans favored aggressive hunter-killer tactics using escort carriers on search and destroy patrols, whereas the British preferred to use their escort carriers to defend the convoys directly. The American view was that defending convoys did little to reduce or contain U-boat numbers, while the British were constrained by having to fight the battle of the Atlantic alone with very limited resources for the early part of the war. There were no spare escorts for sub hunts, and it was only important to neutralize the U-boats which were found in the vicinity of convoys. The survival of convoys was critical, and if a hunt missed its target a convoy of strategic importance could be lost. The British also reasoned that since submarines sought convoys, convoys would be a good place to find submarines.

The Story of VP-63

Once a U-boat was caught and sunk, the chance of the crew getting out alive wasn't good. Source: Deutsche Bundesarchive.

Once America joined the war, the different tactics were complementary, both suppressing the effectiveness of and destroying U-boats. The increase in Allied naval strength allowed both convoy defense and hunter-killer groups to be deployed, and this was reflected in the massive increase in U-Boat sinkings in the latter part of the war. The British developments of ASDIC, Centimetric Radar and the Leigh Light also reached the point of being able to support U-boat hunting towards the end of the war. While at the beginning technology was definitely on the side of the submarine. But Commanders such as F. J. "Johnnie" Walker of the Royal Navy were able to develop integrated

MAD Cats

tactics which made the deployment of hunter-killer groups a practical proposition.

HMS Starling, a sloop of the Modified Black Swan Class, was Captain Frederick John Walker's flagship. In just under a year, this ship sank twelve U-boats. Source: Royal Navy

Eventually "increased allied sub-hunting planes and ships with improved equipment and techniques gradually turned the German submarine service into a suicide organization. Of the thirty thousand men who went to sea in U-boats, two-thirds never returned home." (*PBY: The Flying Boat*, p. 193)

CHAPTER 2

Looking For A Solution

In 1940, the leading scientists in the United States foresaw the impending dangers of the war in Europe. Anticipating the inevitable, Dr. Vannevar Bush, President of the Carnegie Institute, formed the National Defense Research Committee (NRDC) in June, 1940. Bush and his peers in the scientific community felt that they must prepare to make their skills available for the war they knew would come. President Roosevelt chartered the organization with an Executive Order and funded it from his own budget.

Dr. Vannevar Bush. Source: U.S. Library of Congress

One area that the group looked at was antisubmarine warfare. They sponsored a National Academy of Science study of sub-surface warfare, the Colpitts Report, prepared under the leadership of Edwin H. Colpitts. The report was critical of the US Navy's past ASW efforts, and the report was not well received by the Navy. But Bush and Frank B. Jewett, head of Bell Labs, stood behind the report's findings and managed to get Secretary of the Navy Frank Knox to agree to let the NRDC study antisubmarine devices.

Because of Bush and Jewett's persistence, the effort had begun. By January 1941, scientists had surveyed various technologies relevant to antisubmarine warfare and had let contracts for work on acoustic torpedoes and the magnetic detection of submarines. This would become Magnetic Anomaly Detection, which finds a submarine by locating the disturbance a submarine causes in the earth's magnetic field.

MAD Cats

The Bureau of Ships (BUSHIPS) OK'd organizing a new section of Jewett's division of NRDC to consolidate an informal program of work already in progress (*Slide Rule*, p 28). He picked John T. Tate from the University of Minnesota to head the new section, Section C-4. By 18 April, they had published the *Plan for Handling of a Comprehensive Investigation of Submarine Detection*. The group wanted to take a totally comprehensive look at the anti-submarine problem, to study all factors and phenomena involved in the detection of submerged U-boats.

One thing the group began to realize was that progress was uneven. Although the Naval Research Lab had advanced its underwater sound ranging equipment, there was no parallel improvement in the ability to destroy a submarine once it was found. "The ordnance, depth charges, had hardly changed at all; and equipment designed to help the aircraft make submarine attacks was practically nonexistent." (*Slide Rules*, p 28).

A depth charge attack. This picture is of a U.S. Destroyer Escort making a depth charge attack run but the tactics weren't that different when used by either the British or U.S Navies in either World War. The success rate was depressingly low.
Source: U.S. Navy

In June, 1941, John Tate traveled to England to find out what their scientific community had been doing in this area. The British were very cooperative in sharing their experiences with U-boats. They discussed submarine and convoy tactics as well as the performance of their various equipment, including sound gear and depth charges. A major complaint was the tendency of the researchers to continue to work and improve the hardware, never getting systems into production.

The Story of VP-63

On his return from England, Tate managed to get Navy support to set up laboratories on the east and west coasts. On 12 June 1941, NRDC authorized $90,000 for contracts for that month. In July, NRDC authorized another $1,569,500 for additional contracts for antisubmarine warfare research. On 5 September, the Bureau of Ships incorporated Tate's Section into the Navy's official research and development program.

The group continued to look at the problem scientifically, systematically. They managed to discover some facts which led the inquiry to focus on the tasks required for a surface escort to find and kill a U-boat. Theoretically, they discovered, the chance of success was less than 1 in 20. By this point, the scientists discovered that there was one major source of information lacking. They realized that there was a need to quantify the characteristics of the ocean as a medium; they needed to know more about the environment in which a submarine operated.

This opened the possibilities for discovering ways in which to improve existing equipment. More importantly, the group became open to the possibilities of new equipment. The scientists continued to press ahead on a wide front, not always to the pleasure of Navy officials. The results of studies into the fundamental science of acoustics and operational data began to give Tate and his scientists insight into both what was needed in the field and what was technologically achievable. (*Slide Rules*, p. 33) Work continued on a variety of devices, the acoustic torpedo, the echoscope (a short-range sonar), radio sonobouys, and MAD. Work had continued on Magnetic Anomaly Detection equipment, and by October, 1941, had demonstrated a range of 300 feet.

By the and of 1941, the scientists were making a difference. They had turned the Navy's attention from considering only geographical strategies, the positioning of fleets; to studying the U-boat's environment as well as the functional requirements of finding, tracking, and killing a submarine before it could prey on surface convoys. On October 21, 1941, during tests of the newly developing MAD gear, a PBY from the Naval Air Station, Quonset Point, Rhode

MAD Cats

Island, located submarine S-48; proving that the new equipment could actually find U-boats.

*Navy Patrol Bombers did much of the early experimental work.
Source: U.S. Navy*

This effort would eventually come to be called *Project Sail* and formalized as a special airborne testing program for MAD gear. Prompted by promising results, 200 sets of MAD gear would be ordered on July 3rd, 1942. By March, 1942, a Mark IV MAD device in a Navy patrol bomber demonstrated an effective range of 500 feet against a submarine. On June 10th, 1942, *Project Sail* was established to continue the MAD at Quonset Point, Rhode Island. Because of a developing confidence in the technology as a way to locate submarines, over 200 MAD sets were ordered by the Navy.

The Story of VP-63

CHAPTER 3

A Stumbling Start

After Pearl Harbor, Washington was beset with uncertainty and confusion. Admiral Ernest J. King was made Commander in Chief, US Fleet (COMINCH) by President Roosevelt. Admiral King was concerned with getting better tactics and more equipment to throw against the Germans. He was, unfortunately, not terribly interested in exploiting new technology in the area of ASW operations.

Instead of seizing the initiative from the wolfpacks, the US Navy worked from the concept of "convoy vicinity". Surface craft and aircraft, either land based or from escort carriers, would pick off U-boats as they were drawn to the merchant vessels. The idea was to enlarge the defensive area around convoys and providing more escorts. The lessons to be learned from the Royal Air Force's Coastal Command campaign against the U-boat in the Bay of Biscay were not capitalized on. (*Slide Rules*, p. 49)

Admiral Royal E. Ingersoll, Commander in Chief, Atlantic Forces (CINCLANT) was responsible for training all anti-submarine units. He was interested in finding improved techniques for combating the U-boats; but without going up against his boss, Admiral King. So he and his staff quietly explored innovative ideas for defeating the submarine menace. On February 7, 1942, CINCLANT created the Anti-Submarine Warfare Unit in Boston, Massachusetts. Captain Wilder D. Baker was the new unit's chief.

Adm. Ingersoll.
Source: *U.S. Navy*

MAD Cats

During a trip to the UK, Captain Baker was very impressed with the work being done by P.M.S. Blackett and his team; and wanted to develop a similar team in the United States. He was referred to Philip M. Morse, a Massachusetts Institute of Technology (MIT) physicist who was also dismayed at the military's failure to apply science to developing needed capabilities for the war. This was the start of an effort that would develop the needed US advantage in anti-submarine warfare. The new team wanted to develop new weapons, better tactics, and a training program so the users could become adept with the equipment.

The destruction caused along the U.S. East Coast was appalling. For a while, the destruction of merchant ships was so great that there was a serious shortage of food and oil supplies in New England. The U-boat commanders called this "The Happy Time" Source: U.S. Navy

In the beginning, it was frustrating. Through the first half of 1942, German U-boats were devastating. An average of 57 boats were on station in the Atlantic through the fall. They averaged 87 kills per month with only five submarines lost in the same period. The British had lent the US ten corvettes and 24 anti-submarine trawlers and in spite of the best efforts of a wide array of vessels and aircraft, the toll was crushing. Convoys were still not used, and twenty percent of all independent shipping was sunk as the United States lost five percent of its total shipping available. In June alone, 140 ships went down.

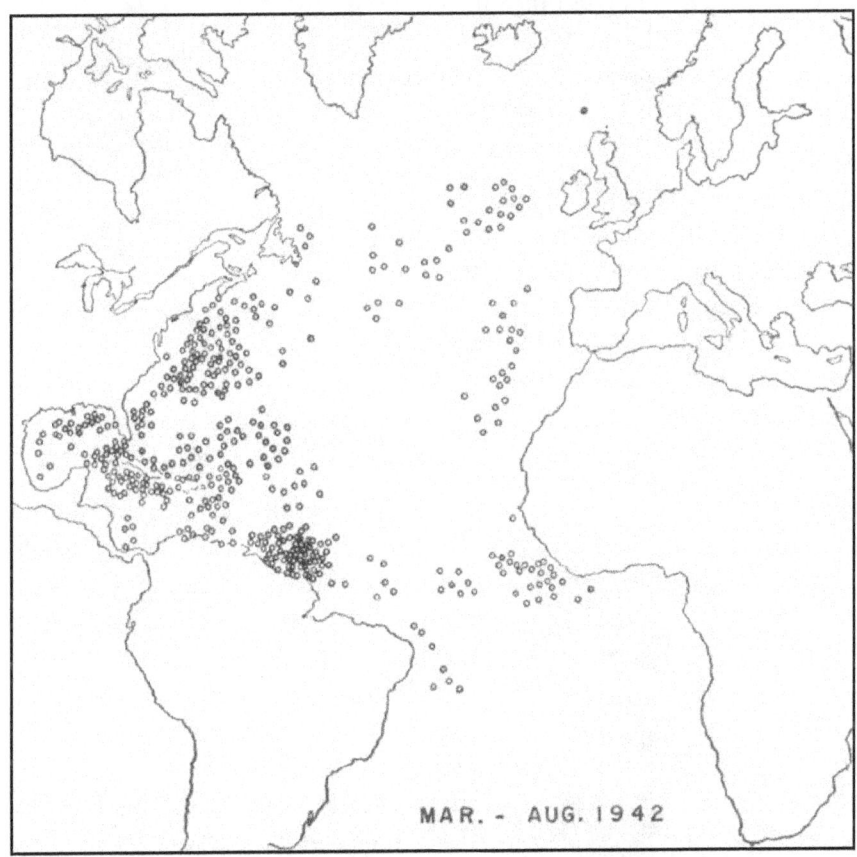

Though the U.S. force could interfere with German operations, it was not until April 14, 1942, that the US drew blood. The destroyer *USS Roper* attacked and sank U-85. By June, East Coast shipping had begun to be convoyed and American anti-submarine work begin to bear some fruit. In April, Admiral King agreed to bring Philip Morse and his team of scientists into COMINCH. One reason for the

MAD Cats

agreement was assurances by Vannevar Bush that COMINCH could control the dissemination of the results of the scientists' work. (*Slide Rules*, p. 53)

But the continuing attitude was that salvation could only come by increasing the protection of shipping. Technological innovation still was not considered the answer. For one thing, the scientists said that a surface attack with depth charges had a five percent chance of killing a submarine.

This was not accepted by the Naval officers. They felt the odds were more like 70 percent (*Slide Rules*, p 54). They did ask the scientific community to help develop training for the new SOund Navigation And Ranging (SONAR) equipment they were fielding. This made it feasible for the group to develop laboratories at New London, Connecticut, and San Diego, California.

First blood was scored by USS Roper, a World War One destroyer modified for escort duties. Source: U.S. Navy

Sonobouys were another part of the puzzle. These are in the bomb bay of a Grumman Avenger. Source: U.S. Navy

They also took advantage of the situation to expand their studies of the underwater habitat. They divided their efforts into three basic categories: location, tracking, and attack. Among the key items they focused on were the ten centimeter radar, improved sonar, and MAD devices. These received the most attention and were felt to offer the best chance for success.

Other efforts included work on sonar improvements, better devices for determining the best time to fire depth charges, and a device prompted by the British; sonobouys. These hydrophone and radio transmitter

The Story of VP-63

devices, the Navy found, could be dropped from an airplane where a submarine was suspected to be submerged. Once the U-boat was located, destroyers could be called by the aircraft and vectored over the boat to attack. The MAD gear and sonobouys would later combine with radar to become the base sensors for VP-63.

The new group working under COMINCH was titled the Anti-Submarine Warfare Operational Research Group (ASWORG). They began to take a scientific approach to collecting data and analyzing the dynamics of an attack on a submarine. In *The Idea Factory,* author Jon Gertner wrote that Philip Morse, an MIT professor who was in Washington organizing (ASWORG) asked Bell Labs to loan their prize physicist, William Shockley, to the organization. "this small think tank ... was staffed by statisticians and physicists and even a chess grandmaster. (*Idea Factory,* p. 71) Because the Navy did not possess reliable data for them to use, the group was able to convince the Navy to allow them to collect data in the field. As a result, scientists joined naval crews in studying their equipment and developing the most promising operating concepts for the new gear.

By mid-1942, a key tactic had become using an airplane to force a submarine to submerge. The plane would then circle the area until the U-boat re-surfaced, using radar to detect it. By re-attacking, the plane would force the submarine to submerge again. This would continue until the U-boat had depleted its batteries and air. A combined, air and sea attack would then sink the sub. This tactic is called hold-down and is still a valuable means of killing diesel-electric submarines.

Secretary of War Henry L. Stimson. Source: Library of Congress

At the same time, Secretary of War Henry L. Stimson was becoming more and more concerned about the U-boat situation. He was dismayed by German successes along the American coast, and about what he perceived as the Navy's failure to aggressively come to grips with the situation. He took his concerns to President Roosevelt,

MAD Cats

who agreed that things were not moving along as rapidly as they needed to. But FDR would not interfere with his Navy Secretary and Commander in Chief.

Unlike others in the chain of command, Secretary Stimson was interested in technology and new weapons. So he would eventually become the catalyst in rallying forces and focusing development in new areas. Vannevar Bush presented Stimson with the argument that the airplane and radar could go a long way in defeating the submarine. After observing first hand the operation of radar, he began to push the new approach, a push that would open the door to accepting innovation.

Capitalizing on the British experience, Stimson ordered the Army Air Forces to develop a unit to operate against U-boats. The Sea-Search Attack Development Unit (SADU) was formed at Langley field, Virginia, in May 1942. Edward Bowles, Stimson's special consultant for radar, equipped a fleet of 90 long-range bombers with ASV-10 ten centimeter radars. In spite of some organizational and operational problems with a rigid Navy command system, the Army planes began operations, successfully pressing attacks in July, August, and September of that year.

The B-18 was obsolete as a bomber but with an ASV-10 in the nose and a MAD unit in the tail, they were a useful interim ASW aircraft.
Source: U.S. Army Air Force

Even though progress was being made in developing ideas for better use of the airplane in conjunction with surface vessels to counter the submarine threat, scientists pushed for a central authority to control the seeding of prototype development, coordinate the availability of aircraft and equipment, as well as manage crew training. But Navy brass was still rigid in its thinking that convoys were the only way to protect merchant shipping.

The Story of VP-63

By September of 1942, convoy tactics in the western Atlantic and Caribbean took away the U-boats easy pickings in those areas. So Doenitz shifted his force, which averaged 100 boats at sea, to the northern convoy lines in mid-ocean. U.S. tactics and effectiveness had not improved significantly. The initiative was still with the Germans and the Americans continued to rely on tactics that were about as effective in 1942 as they had been at the end of World War I.

While aircraft lacked a means to track submerged submarines, a successful attack meant overwhelming the submarine before it could dive. Aircrews would throw everything they had, bombs, rockets, cannon and machinegun fire, in a desperate effort to cripple the submarine before it could get to safety. Source: Royal Air Force

Finally, in the winter and spring 1942/43, the US Navy anti-submarine effort began of be consolidated. The efforts of the scientists that had begun in 1940 were on the verge of deployment as weapons. At long last, it seemed that we were ready to effectively take on the U-boats. But the U-boat threat was increasing dramatically. Deonitz was able to deploy 70 to 80 boats to the Atlantic and extend their patrol time. He had also begun hitting the convoy routes to and from England hard. His *Rudeltaktik* (wolfpack tactics) had been developed to near perfection. He also was keeping his boats out of reach of the shore-based aircraft, concentrating on materiel-laden ships en-route to England in the western Atlantic and ships returning to the US in the eastern Atlantic. What was needed was a way to keep the U-boats from getting to where they could attack Allied shipping, and idea that would eventually focus on where the submarines entered and left the ocean to begin or return from patrols.

An important die was being cast, however. Allied use of airplanes was threatening to make life tough for the U-boats. Between July and

MAD Cats

September 1942, RAF planes operating over the Bay of Biscay were sighting every second U-boat crossing the Bay of Biscay, sinking five. This was finally taking on the U-boats before they could become a threat to the supply line.

In the North Atlantic, though, things continued to be bad. Airplanes could only cover a limited area and bad weather increasingly favored the submarine, plus there was a shortage of escorts. But it could have been worse. The U-boats were spread thin and finding a convoy often depended on luck. Many convoys were able to get through unscathed. In 1942, the Allies ran 246 convoys averaging 31 ships. 127 ships were lost.

The British found the Liberator to be an excellent long-range maritime patrol aircraft from March 1941 onwards. The aircraft above is one of the first to enter British service and has an odd mixture of American markings and British serial numbers. Source: U.S. Air Force.

Britain was facing major problems due to the German attacks on its all-important lifeline. But there was a major conflict between the US and British position as to how to approach the problem. Admiral King continued to hold to the strong convoy escort concept while the British

The Story of VP-63

held that U-boats should be attacked at every turn. The British felt that their Bay of Biscay campaign was bearing fruit and they lobbied for more B-24 Very Long Range aircraft.

At the Atlantic Convoy Conference in Washington during March, 1943 the British also pushed for a joint, coordinated anti-submarine effort, one with more unity of command. They wanted a coordinated development of ASW technology as well and were supported by the American scientists.

While Admiral King never fully supported the changes proposed and did not want to relinquish control over the Navy's anti-submarine forces; he had to do something to ward off the threat of intervention by the President. On April 6th, 1943, he appointed Rear Admiral F.S. Low as Assistant Chief of Staff, Anti-submarine Warfare. Low was convinced that the Navy's anti-submarine warfare organization was flawed and made the recommendations that the Navy take a more aggressive approach to combating submarines and begin to install the special devices the scientists had been

Admiral Ernest J. King.
Source: U.S. Navy

working on. This included Magnetic Anomaly Detection and the new retro bombs developed to go along with them.

On May 20th, 1943, King established the Tenth Fleet to assume responsibility for the American anti-submarine effort. Officially, Admiral King still commanded Tenth fleet because he kept it in COMINCH. But Admiral Low actually ran the organization. It had been seventeen months since hostilities began. And at last, due primarily to political pressure from the Secretary of War and the

MAD Cats

British, the Navy finally had an organization responsible for taking the initiative against the U-boat. The weapons to do it with were at hand; and so was VP-63.

VP-63's Catalinas arrived at the opportune moment. Source: U.S. Navy

The Story of VP-63

CHAPTER 4

A Good Plan Comes Together

In early 1943, the Navy was pressing forward with programs to counter the German U-boat menace. No longer fettered by old, constricted thinking, the scientists and engineers were working with military personnel to bring into operation systems that would prove to be a decisive factor in deflating the submarine problem. Efforts focused on improving the ability to defend convoys as well as finding ways to maximize the utility of the airplane. Land and sea-based high frequency direction-finding equipment began to warn of the location of German wolfpacks. Radar detected the U-boats as they moved in for attack. By early 1943, radar detection range had gone from 2,700 meters (1.67 miles) to 7,000 meters (4.3 miles). Better sonar was effectively providing a close-in means of keeping track of the boats and guiding the final attack.

In his classic fifteen-volume *History of the United States Navy,* in *Volume X on the Battle of the Atlantic,* Samuel Eliot Morrison describes the birth of an effective effort against German submarines. All U-Boats had to transit the Bay of Biscay twice every war patrol, Air Vice Marshal Sir John Slessor sent a 20 April, 1943, memorandum to the Anti-U-Boat Subcommittee calling the Bay the trunk of the U-Boat menace tree, with the roots being the Biscay ports. "The tree should be felled by severing its trunk – the little patch of water 300 miles by 120 in the Bay of Biscay, through which five out of six U-Boats operating in the Atlantic must pass." (*History* pg. 86).

Tactics began to press for offensive operations against the critical operational nodes of the U-boat system. The idea was now to attack the submarines before they could get to the convoys. The ASW

MAD Cats

forces would create a deadly gauntlet for the boats to cross; to bleed them as they crossed the Bay of Biscay. (*History*, pg. 88)

The German Metox receiver was effective but a shot-down allied pilot invented a story about how allied aircraft could detect its emissions. The shocked Germans tested Metox and found it did give off weak emissions. They then withdrew it. Source: Deutsche Bundesarchive

From March to mid-July, Coastal British Forces were carrying the whole load. One of the factors that countered the February offensive was the German radar receiver *Metox*, which enabled U-boat skippers to detect Allied planes using the best radar available at the time. In the Spring of 1942, the Germans had captured an ASV Mark II aircraft radar, using the information from the system to develop the *Metox* R-600 receiver so the boats could detect aircraft searching for them in the new offensive. The systems proved to be unreliable and difficult to maintain; but by November the entire fleet had them and were using the warning device to try and avoid detection.

But on 20 March, 1943, the British Coastal Command uncorked the ASV-10, the 10 cm-wave radar on which American and British scientists had been working. The radars plus 80,000,000 candle-power Leigh search lights on RAF Wellington bombers made it possible to attack U-boats at night from the air with great success. As soon as a surfaced U-boat was picked up on ASV, the plane made a radar approach undetected, turned on its blinding light at a few hundred yards, and attacked before the boat even knew it had been sighted. (*History*, p 89) The campaign got under way on 22 March, 1943, when a Wellington bomber sank U-665 at night about 280 miles West South West of Brest, in the Bay of Biscay patrol area.

The Story of VP-63

A Leigh Light fitted to an RAF Liberator. The blinding searchlight made night attacks on surfaced U-boats an extremely effective tactic. Source: Royal Air Force.

MAD Cats

Admiral Doenitz could not get the Luftwaffe to provide his boats adequate air cover over the Bay of Biscay. Until late August U-boats in transit were given very little protection by the Luftwaffe, whose capabilities, with several bases near the coast of France, were very great. The reason, apparently, was the feud between Goering and Doenitz. The portly *Reichsmarschall* did not care to build up the *Grossadmiral's* prestige at his own expense.

Reichsmarschall Goering is often criticized for his failure to put fighters over the Bay of Biscay to cover U-boats. However, the problem was more complex than his critics realize. German fighters were very short-ranged and putting effective patrols over the Bay would have required an inordinate number of aircraft. The only fighter with enough range for such operations was the Ju-88C3 shown here. A converted bomber, it was dead meat when faced with the heavily-armed and armored British Beaufighters that prowled the Bay looking for them. Source: Deutsche Bundesarchive,

The very thin fighter and bomber patrol that he put on over the Bay of Biscay accounted for only one victim, a MAD Cat. This was a *Catalina* nicknamed *Aunt Minnie* piloted by Lieutenant William P.

The Story of VP-63

Tanner USNR, who as pilot of a PBY patrolling off Pearl Harbor had sighted a Japanese midget submarine on the morning of 7 December 1941. Unfortunate *Minnie* was attacked by 8 to 12 Ju-88s on 1 August. She splashed one on the first pass, then came under a fatal crossfire, burst into flames from wing to wing, and had to be ditched. Tanner with his co-pilot and waist-gunner climbed on board a life raft and were rescued, after 24 hours, by HMS *Bideford*.

A struggle for aircraft began. Sir John made repeated efforts to obtain more American B-24 Liberators. Air Marshal Slessor wanted to take full advantage of U-boat targets in daylight. His force was bolstered with British Sunderlands and Halifaxes. Overall, the May offensive logged 98 sightings on 112 boats transiting the Bay. The Coastal Command attacked 64, sank seven and damaged seven more. Six aircraft were lost.

The German Navy modified seven U-boats with greatly-increased anti-aircraft armaments to escort other U-boats across the Bay. The idea proved to be a very bad one and the survivors were converted back to standard. Source: Deutsche Bundesarchive.

Then Doenitz tried new tactics. On June 1st he ordered all boats to make the transit in groups, surfaced. Five U-boats transited inbound and two or three outbound. He hoped that the combined flak could drive off the aircraft. This tactic worked for a while; the first two pairs made Brest safely on June 7th and 11th. On 12 June, three unarmed British aircraft detected five boats in company about 90 miles north of Cape Ortegal. The three planes circled the area for three hours unsuccessfully hoping to attract armed help. The boats submerged and made a night run to clear the Bay. They were spotted the next day 250 miles west of Finisterre by a long-range Sunderland and the tactic was uncovered. (*History*, p. 92)

MAD Cats

The *Sunderland* attacked, damaging U-564, rupturing its pressure hull with bombs but taking flak hits itself. Three subs continued to the open sea, while U-185 turned to escort the U-boat that had fallen victim to the *Sunderland* back home. They were attacked the following day and U-564 sent to the bottom. Also that day, a group of five and a group of two were discovered transiting north of Cape Ortegal. They were attacked and two boats damaged. That night, Doenitz issued orders that groups of U-boats would proceed through Biscay mainly submerged and surface only to charge batteries.

Encouraged by the Bay operation, Admiral King agreed to give Sir John six squadrons of *Liberators*. On 7 July, the first two squadrons of American B-24s, 24 planes, were added to the Bay operation. They flew their first operational missions July 13th. U-boats were now entering and leaving in groups, and the planes had good hunting. Coastal Command planes had sunk five boats since the 1st of July. On the morning of the 20th a B-24 bombed U-558 off the mouth of the Bay. The crew abandoned ship while a British Halifax and a second Liberator finished up the boat.

The Coastal Command was encountering transiting groups on the surface; with radar at night, visually during the day. Doenitz hoped that boats could provide mutual anti-aircraft support and protection. Once detected, boats turned tight circles firing AA guns. At some point, a skipper would lose his nerve and straighten out to dive. The planes would begin their bombing run.

In July, the US Navy got into the Biscay show. At Slesser's request, Admiral King sent VP-63 over in the last week of July. The unit departed Reykjavik 20 July, 1943, for Pembroke Dock, South Wales. It began patrols on 25 July. The Army retired from ASW and was replaced by the Navy Liberators. At the end of September, the Navy had 44 combat aircraft, (30 Liberators & 14 Catalinas) under Coastal Command.

VP-63 detachments had seen duty at various times from Elizabeth City, New Jersey; Quonset Point Rhode Island, Jacksonville and Key West Florida; as well as Naval Air Station, Bermuda. On 7 June, the squadron was ordered to Iceland; but found no submarines.

The Story of VP-63

In *PBY: The Catalina Flying Boat* Roscoe Creed explained that "even though the US Navy PBYs in Iceland (and Newfoundland) sank few U-boats, they did the next best thing; they stopped them from sinking Allied ships." Their tactics disrupted the U-boats' attack patterns, rendering them more and more ineffective. It was after wolf pack tactics peaked that VP-63 worked its way across the Atlantic to Pembroke Dock in South Wales; with duty stints in Argentia, Newfoundland and Iceland. (*PBY*, p.193)

They were assigned to the 19th Group, RAF Coastal Command. During that six-month tour, VP-63 flew more hours and missions than any other squadron attached to 19th Group. Although they were not responsible for any findings or sinkings, mostly because the area was not favorable for MAD operations, the squadron polished its tactics and equipment.

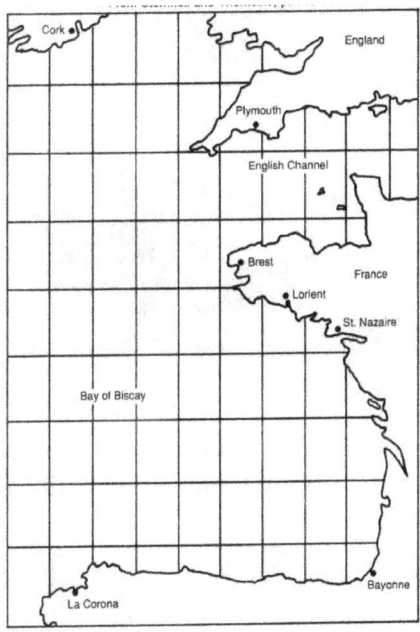

In December, 1943, VP-63 was moved to Port Lyautey, French Morocco. By operating in the narrow waters of the Strait of Gibraltar, VP-63 found that the narrow confines favored the MAD technique. A 'magnetic barrier' across the Strait was the result of

MAD Cats

two aircraft flying a race-track course four miles long and three-fourths of a mile apart over the deepest part of the passage, on a line running from Pt. Camarinal, Spain, to Pt. Malabata, Spanish Morocco. They would fly at altitudes from 50 to 150 feet at positions always opposite each other. As a result, an aircraft would pass any point on the circuit at one to two minute intervals. Supposedly, no U-boat could pass without being detected by the range-limited MAD gear. Two shifts covered "The Fence" from dawn to dusk. (*PBY*, p. 208-213)

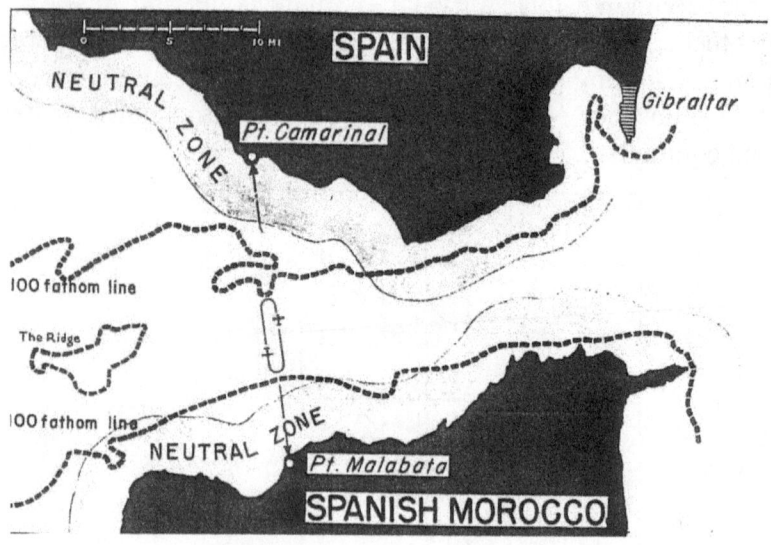

Creed credits VP-63 with being in the right place at the right time, and key to three of the most important and enduring anti-submarine technologies that endure today.

Although the Germans claimed that the MAD Cats had been chased away, they stuck to it until the end of 1943; and by that time had carried out more Bay patrols than any other squadron in Air Vice Marshal Bromet's 19th Group.

The Story of VP-63

CHAPTER 5
The Cats Draw Blood

For Lieutenant Commander Edwin Wagner, and the crews of VP-63, 1943 had been a frustrating year. The unit had received its MAD-fitted *Catalinas* at the close of 1942. Early in 1943 the squadron's aircraft began operating off the east coast of the USA and, later, off Iceland; but on each occasion the arrival of VP-63 seemed to coincide with a reduction in U-boat activity. Some people even joked that the MAD Cat Squadron was a solution in search of a problem.

In July the unit moved to Pembroke Dock in Wales to take part in the Battle of the Bay; but while the special detector and the retro-bombs were effective against U-boats cruising underwater, the slow-flying PBY MAD Cat itself was vulnerable if the submarine stayed on the surface and fought back. Patron 63 began to patrol the Bay on 25 July. Based at Pembroke Dock in South Wales, Patron 63 began to patrol the Bay on 25 July. Two Catalinas flushed U-262 and U-760 about 150 miles NW of Finisterre on the 28th and engaged them at medium range while awaiting air reinforcement; but the boats escaped. The MAD proved to be of little or no use in the Bay.

So after a short time VP-63 left the area, having again achieved little, as far as sinking submarines went. The squadron, now commanded by Lieutenant Commander C. Hutchings, received orders for a further move; to Port Lyautey in Morocco. to reinforce the units covering the Straits of Gibraltar.

Since the October disaster, the German Navy had had some success in sending U-boats through the Straits. One made the passage it November, two in December and four in January—all the U-boats attempting the passage had in fact made it safely through.

By their very geography, the Straits of Gibraltar are one of the finest places in the world in which to conduct a MAD search for

MAD Cats

submarines. The waters are narrow: at its narrowest point the deep-water sleeve is only four miles wide. The waters are deep: the 'sleeve' is rarely less than 600 feet deep except close to the shore, so there is little or no magnetic interference from wrecks lying on the bottom. The currents flowing through the Strait are complex, and always subject to large fluctuations; but as a rough generalization the water down to the first 100 to 150 feet flows eastwards into the Mediterranean, while the deeper water flows westwards into the Atlantic. At its strongest, the deep westward-flowing current reaches 4 ½ knots—sufficient to render a deep passage by an eastbound U-boat virtually impossible. That the eastbound German submarines were unlikely to be deeper than 150 feet was important to the crews of the American MAD Cats, for their equipment could not register U-boats more than 400 feet away. Moreover the optimum German tactics, to approach the Straits charging the batteries during the night and run through submerged during the day, also suited the MAD Cat crews well: their entire patrol had to be flown at a very low level, and this was possible only by day.

A MAD Cat operating from Port Lyautey
Source: U.S. Navy

Before VP-63 began to patrol the Straits of Gibraltar, the MAD had been considered mainly as a follow-up device, to re-locate boats which had previously been seen on radar; but which had submerged. Such were the limitations imposed on submarine navigation by the geography of the Straits, however, that it was now possible to use the MAD as a primary detection device. Mr. John Pellam, of the US Operational Research Group, devised the tactics which were soon to be employed to good effect. The aircraft were to fly a 'barrier patrol', round an oblong four miles long and one mile wide,

The Story of VP-63

lying astride the narrowest point in the deep water channel. Two MAD Cats, maintaining position opposite each other on the oblong, were to fly round it at a true airspeed of 115 mph; thus any given point on the oblong was to be passed once every three minutes. Against a 200-foot long submarine, moving at two knots through a favorable current of two knots, this meant that there was a very good chance of the boat registering on the MAD equipment twice as it passed through the barrier.

The MAD Cats of VP-63 established their barrier across the Straits during February 1944. Every morning, all morning, two aircraft flew up and down the oblong, completing one orbit every six minutes. Each noon they were relieved by a second pair, which repeated the performance during the afternoon. To get the best possible range on underwater objects, the MAD Cats had to fly at an altitude of no more than 100 feet above the water. This took away some of the monotony; as one of the American pilots later recalled performing turns and wing-overs close to the water. It was even legal!" (*Aircraft versus Submarines*, p. 194)

* * * * *

VP-63 drew first blood when U-176 crossed the MAD Cat barrier twelve days after leaving port in Brest. The encounter was described by squadron members, a Navy Press Release, and summarized by Alfred Price in *Aircraft versus Submarines*.

Shortly before 1600 hours on the 24th of February, Aircraft Radioman J. Cunningham with Crew 15 was operating the MAD equipment. He reported that the recording pen in front of him had suddenly zig-zagged across the paper trace, signaling that there was a magnetic disturbance below. The pilot, Lieutenant (jg) T. Wooley, began the often-practiced 'clover leaf' tracking pattern. The second MAD Cat, piloted by Lieutenant H. Baker, joined in. They established that the source of the magnetic disturbance was in fact moving and made repeated runs over the area, releasing a smoke marker each time there was a MAD indication, producing a line of

MAD Cats

markers on an underwater metallic object moving eastwards at about two knots. They were certain this was a submarine.

The destroyer HMS *Anthony* was in the area patrolling the Straits and attempted to engage. *Anthony* made contact on her ASDIC; but conditions were very poor and contact was soon lost. The destroyer's presence forced the PBY's, which operated at VERY low altitude, to climb away to avoid a collision with the ship. They lost MAD contact as well.

The MAD Cats restarted their search, flying in a wide circle centered on the point where they had lost contact. Half an hour later, at 1645 hours, the pen recorder in Wooley's aircraft zigzagged. Once again Wooley and Baker went into their tracking procedure, and laid out a line of ten smoke markers to indicate the submarine's course and speed. They radioed a rather pointed request to the destroyers to remain in the area, but to keep clear of the smoke floats.

Finally, Lt. Wooley judged it was time to attack with a full complement of retro-bombs. At 1656 hours, when the MAD indicated "overhead", he pressed the button to fire the bombs. One failed to ignite, but the remaining twenty-three roared backwards off their rails, fell vertically, and entered the water in a rectangular pattern.

According to after-action interviews, the crew of U-761, unaware of the activity above, had been proceeding submerged at a depth of 50 meters. The layers of water had different densities, making ASDIC conditions difficult for the operators on board HMS *Anthony*; but also making it difficult for the U-boat to stay at a predetermined depth. Then, with no warning, there were four loud detonations against the submarine's hull. The German sailors could only speculate on the cause.

As the reports from the various sections of the boat reached Oberleutnant Geider, it appeared that the U-boat had suffered no major damage. But one thing was clear, their attempt to sneak through the Straits had been discovered. Price wrote that "The

The Story of VP-63

feelings of those on board U-761 were rather like those of a burglar creeping through a darkened house, who has just felt a stone bounce off his shoulder."

Two minutes later, Lt. Baker's aircraft attacked with all twenty-four of his retro-bombs. Then *Anthony* put down a pattern of ten depth charges, across the head of the smoke markers. After this U-761, was in serious trouble. She reared up to the surface out of control, stern down and with little way on, sank back stern first. The destroyer HMS *Wishart*, also in the area, made contact and laid another pattern of depth charges, followed by another attack from the *Anthony*.

Damage to the German submarine was grave. All electrical installations including the batteries and motors were damaged and out of action. The main electrical control panel was destroyed, the hydrophones and all the radio equipment was smashed, and the compressors had been wrenched loose from their mountings. Water had entered the boat through a loose valve in the drain pump and the clutch between the diesels and the motors was jammed Some of the high pressure air lines had been ruptured and all lighting except the emergency lights was out. The air within the boat had become quite foul, and there were alarming signs of chlorine gas. Oberleutnant Geider ordered the boat to surface for the last time As she did so she came under attack from an RAF *Catalina* and a US Navy *Ventura*. The sailors struggled clear of the sinking sub, and destroyers picked up forty-one of them.

The second success was described as follows:

The next U-boat to attempt to pass the Straits was U-618 but Kapitanleutnant Baberg found the defenses too strong and returned to France. Oberleutnant Schuemann in U-392 was more determined and tried to sneak through on 16 March. The MAD Cat sentries were flying their barrier tracks and the U-boat was detected almost immediately. Based on the lessons of the previous operation, this time the co-operation between ships and aircraft was almost perfect. The destroyers stayed clear, while the MAD Cats traced the U-boat with smoke markers and attacked with retrobombs, scoring three

MAD Cats

hits each. The senior aircraft commander then called in HMS *Vanoc*, which made an unsuccessful attack with her anti-submarine mortar. Finally HMS *Affleck* finished off the still-submerged U-392 with three hits from her mortar.

During the remainder of March three U-boats passed successfully through the Straits by sticking to the shallower water on the southern side. But on 5 May, the MAD Cats located U-731. The action was almost a repeat of that which sank U-392. Only one more U-boat attempted the dangerous Straits passage, and she cleared the defenses successfully on 17 May. After that, events would force the German Navy to concentrate its attentions elsewhere. (*Aircraft versus Submarines*, p. 194-196)

Part 2

The MAD Cats

A History of VP-63

The Story of VP-63

The MAD Cats - The Story of Patrol Squadron 63

The following is from the unpublished "The History and Tales of the MAD Cats" and was written by the unit's two wartime skippers, Commander E.O. Wagner and Lieutenant Commander C.H. Hutchings.

How did we become MAD Cats in the first place? Situation: In 1940, the leading scientists of the United States foresaw the impending dangers of the war in Europe. They accurately assessed the technical shortcomings of our Army and Navy and decided to do something about them. They formed the National Defense Research Committee (NDRC). This organization, with Presidential support and direction, was to contribute many war-winning developments in all combat areas including air and anti-air warfare, submarine and anti-submarine warfare, nuclear warfare, Command, Control and Communications, to name a few. The VP-63 MAD Cats were created as a result of the work directed by the National Defense Research Council and its supporting

MAD Cats

organizations, universities, laboratories, and teams. We MAD Cats were born out of United States science.

Initially we were commissioned as a regular Air Force Pacific Fleet Navy patrol squadron. Our orders were to form, organize, train, and proceed to Guadalcanal as quickly as possible. While we were doing this with PBY-5As, scientists from the California Institute of Technology were perfecting retrofiring rockets and testing them at a range at Goldstone Lake, California. This work was in conjunction with the Magnetic Anomaly Detector being developed by Columbia University scientists. These joint tests showed promise.

At the same time, German U-boats continued sinking ships along the Atlantic Coast at an alarming rate. The available anti-submarine forces did not appear to be able to cope with the situation. New approaches were needed. The Commander-in-Chief of the U.S. Fleet ordered that a patrol squadron be designated to give the MAD retro-rockets a try. And that is how our anti-submarine history was to begin.

There we were, a mixture of American kids of varying experience and education, brought together to do a very sticky job. We had to organize and train the basic aircraft squadron; design installations; install the retro-rockets, retro-float lights, and MAD equipment; develop maintenance and operating procedures; and develop tactical procedures for searching, detecting, and attacking submerged submarines. As a matter of fact, very few people ever dreamed of an airplane tracking and attacking a submerged submarine. This was our first challenge. It led to the adoption of our motto: "The extremely difficult we do immediately; the damned impossible takes a little longer, but we do it just the same!" Somehow we knew that we were destined for something very special.

The complete help of the scientists, of course, was necessary in all of the technical areas. Without them we couldn't have done it. We used to call them the "confidential professors" because of the high classification of their "hush-hush" equipment and concepts. We also remember that we were never without one of their representatives. In fact, they were the finest shipmates.

The Story of VP-63

As for ourselves, it was a sailor's job not only to learn the installation and the equipment, but also to fly it under a wide variety of conditions, to maintain it, and to operate it under circumstances that only occur in a patrol plane with its crew, alone at sea.

Some of us remember that in a small way we started preliminary anti-submarine work in October 1942 using PBY-5As. Then we had to ferry and turn over these planes to Pearl Harbor and begin over again with new PBY-5 seaplanes which we drew out in January 1943. We conducted training in NAS Alameda (California) using motor launches in San Francisco Bay to practice and experiment with spirals, trapping circles, and any other ideas we could think of. We also used an armored target boat in Monterrey Bay where we practiced trapping circles and tracking. (We took advantage of) the opportunities to operate in cooperation with submarines leaving Mare Island Navy Yard after their overhauls. We practiced glide bombing runs at Sunnyvale, (California). We were getting started. Then we borrowed an S-boat submarine from the Sound School at San Diego. We flew and argued until we hammered out the early basic tactics. We were now ready for concentrated training.

We had demonstrated our capabilities to detect, track, and attack submerged submarines to Commander Air Force Pacific and he was sold. Soon after, however, the Commander-in-Chief, U.S. Fleet was having second thoughts. With ComAirPac's support, we made a pitch in Washington which apparently worked because we not only got our final go-ahead, but our complement was increased to 18 planes and crew. We were now ready to roll.

With this approval and support we were able to get a target submarine for four straight weeks. We flew and flew until all of our patrol plane commanders "got hot". Early in this period we started to use the miniature retro-rockets. We remember that these were armed with shotgun shells. On the first hits, the sub skipper was certain that we were trying to do him in with depth charges. Some cork insulation was loosened and the noise of the shotgun shells exploding against the hull was too realistic for comfort. But it helped us make our point.

MAD Cats

On March 15, 1943, we left the Pacific Fleet to report to our new bosses in the Atlantic. We were on the way! What was the Atlantic Fleet getting? They were getting a new type of pioneer. We pioneered by being the first aircraft squadron ever to be equipped with rockets and we were the first squadron to be equipped with Magnetic Anomaly Detectors. Incidentally, we were also the first squadron to use sonobouys, aircraft radar, radar receivers, and absolute altimeters. With all of these new systems, we had to be virtually self-supporting. We were now "hot" and "ready"! For the record, we can say of ourselves that the magnitude and complexities of our many changing tasks ranged somewhere between the extremely difficult and the damned impossible. But we accomplished them just the same - and in less than six months! The MAD Cat motto was working!

After a short period at Elizabeth City, North Carolina, we were moved to NAS Quonset Point, (Rhode Island) where we became an operating unit of the Anti-submarine Development Detachment and of the Atlantic Sea Frontier. Strategically, this appeared to be a good situation for us.

Soon there was a submarine reported to be in the Key West area. We sent a detachment immediately. They flew and flew - no contacts! Then there was a submarine reported to be in the Bermuda area. We sent a detachment immediately - no contacts! If they didn't know it before, it was now plain that the German U-boats had left the entire Atlantic seaboard area. Where did they go?

Lt. Cmdr. C.H. Hutchings, the Executive Officer of the squadron, reported that one of our aircraft on patrol out of Bermuda flew through the eye of a hurricane - a harrowing experience - but the base aerologist was delighted to learn that a hurricane was there!

One night at Bermuda, a *Coronado* four-engine seaplane from another squadron failed to return from patrol. Wing operations personnel concluded (based on radio bearings) that the plane was down somewhere southwest of Bermuda. However, our analysis of the flight assigned, the weather, and other factors led us to believe that the *Coronado* was down northeast of Bermuda. In spite of the

The Story of VP-63

fact that it was night and raining, with a ceiling of 1000 feet and a 40-know wind blowing, five of our MAD Cat aircraft departed on a search and rescue mission to the northeast. The first aircraft off the water flew out an estimated 300 nautical miles and threw out a float light to maintain position and request further instructions.

Then the impossible happened. A Very Star was fired from the surface less than 50 yards from the float light. The crew of the *Coronado*, in two life rafts, had been found. We maintained contact with the life rafts until the *Coronado* crew was picked up by a destroyer about noon the next day. What are the chances of our MAD Cat crew almost hitting the life rafts with a float light under the conditions which prevailed? One in a thousand! Was the pilot guided by God? We don't know, but this we do know - if we had not been willing to undertake the impossible, the successful rescue would not have happened. Incidentally, a few days later one of our planes on patrol landed on the open ocean and rescued three British sailors who had been adrift in a raft for about 40 days.

After our operations all up and down the Atlantic Coast, the Commander-in-Chief ordered us to Iceland. We flew and we flew in the region of the Arctic midnight sun - no contacts! What was left? The United Kingdom!

Admiral King deployed us to Wales and loaned our squadron to the Royal Air Force in response to the request of Air Marshall Fenner for their big Bay of Biscay Offensive. Now for sure we were going to get to show our "stuff"! Not so! We all remember that the first thing that happened was that one of our planes was shot down by a large flight of Ju-88s. This was a sobering beginning.

The British credited us in this action with shooting down two Ju-88s and damaging a third so severely that it could not return to base. This assessment was based on intercepted Ju-88 messages. (Subsequently a U.S. Navy Department press release dated July 7, 1945, stated that we "shot down two Ju-88s in an engagement that marked the first aerial meeting between U.S. Naval Aviation and the Luftwaffe..."). One of the MAD Cat survivors of this action was

MAD Cats

reassigned to VP-63 at his own request after he recovered from his critical injuries.

On one of our first patrols down into the Bay, one of our planes spotted two U-boats in close formation on the surface, but the U-boats crash dived and our plane could not get to the point of submergence quickly enough to contain the U-boats in a trapping circle. The U-boats escaped. This was a big disappointment, but our spirits remained high.

In July, 1943, some U-boats adopted a policy of staying on the surface and "shooting it out" when against a single aircraft. To counter this tactic, our MAD Cat personnel designed and tested fixed forward-firing .50 caliber machine guns mounted on the wings. The system was highly effective; but was never used, because the U-boats resumed their old tactics of diving whenever approaching aircraft were detected.

Then we discovered that the rocket bomb racks made for us by the Bureau of Ordinance were causing structural failures in the wings of our aircraft. So our personnel designed, made, tested, and installed radically different racks on all aircraft. The new racks enabled us to carry 30 bombs per plane (6 more than previously carried) without any increase in weight. Moreover, we experienced no further wing failures.

At Pembroke Dock we received dispatch orders from the Bureau of Naval Personnel to transfer 10 of our 12 MAD technicians. Since qualified replacements did not exist nowhere in the entire U.S. Navy and our men wanted to stay with VP-63, we vigorously protested the orders. After several exchanges of dispatches, BUPERS canceled the orders.

Bureau of Aeronautics instructions specified that the maximum gross weight of *Catalinas* should be no more than 35,000 pounds, yet our aircraft when fully loaded weighed 60,000 pounds. Everyone except our personnel said this was impossible; yet we did it. Admittedly, take-offs took a full minute and we had to "hang on the props" for about the first two hours of flight.

The Story of VP-63

Losses of Coastal Command patrol planes to Ju-88 aircraft ranged from two to four percent - one aircraft out of twenty-five patrolling the Bay of Biscay failed to return from patrol. Yet VP-63 lost only one aircraft while flying 449 patrol missions, even though flights of Ju-88s were sighted on at least 15 occasions. Our tactics were simple - detect the Ju-88s visually or with radar as far away as possible, then head west at full throttle and run for the clouds. We had trained with RAF Spitfire fighters using camera guns in case we had to fight.

We remember those time-consuming flights from Land's End to the northern coast of Spain. The Bay of Biscay operation was a winner. We out-flew, in missions per plane, our RAF brothers by a margin of about three to one. Our crews and the PBY-5s were in a class by themselves.

Our accomplishments in the Bay of Biscay did not go unnoticed as indicated by some of the messages we received. To cite a few: "COMFAIRWING 7 noted with pleasure the splendid record of operations of PATRON 63 while in the area. During the period of operations from 12 July to 14 December 1943, PATRON 63 flew more hours in operations against the enemy than any other squadron regardless of size attached to 19 Group Coastal Command RAP. During the months of August, September, and October, PATRON 63 was high squadron in number of sorties as well as hours among the flying boat squadrons attached to 19 Group. Suitable recognition should be instituted for all individuals concerned. "Well done and good luck."

A message from Commander Naval Forces in Europe states: "The performance of PATRON 63 while engaged in the Bay of Biscay Offensive brought honor to the Navy and prestige to the squadron."

A message from the Commander-in-Chief and Chief of Naval Operations stated: "The Commander 10th Fleet notes with pleasure and approval your squadron record of operations in the Bay Offensive." The grateful residents of Pembroke Dock honored VP-63 by depicting one of our MAD Cat aircraft on a stained glass window in one of their churches.

MAD Cats

An assessment at this point highlighted a few interesting aspects of our squadron. We had flown thousands and thousands of operational miles. Our flight crews and our maintenance teams performed beyond all expectations. We lost no airplanes due to engine failures and had no airplane system failures. Especially interesting was the outstanding state of morale which was maintained under a wide variety of living and operating conditions. There were no disciplinary actions requiring a Captain's Mast.

Of course there had to be an explanation of why we couldn't find and attack German submarines after that much flying. A few brief facts are in order. American technology put the fear of God into the crews of the German subs. No ships were attacked by U-boats while they were covered by PATRON 63 ASW air coverage. The Bay of Biscay Offensive strained the endurance of the German subs to traverse the covered areas. Because this exhausted their batteries, they were forced to surface or snorkel often to recharge them. Thus they were exposed to our carrier-based shipmates who also had a good system which they used well.

In general, it might be said that at this point the ASW campaign essentially was won. The German high command was forced to make major strategic changes in their plans. But for us, it was finally just beginning.

Closing the Mediterranean Sea to German subs remained a problem. Maybe we could solve it. So Admiral King ordered us to move from Pembroke Dock, Wales, to Port Lyautey, Morocco. Life for us was to take on an added significance.

The Story of VP-63

After being sighted by a MAD Cat, German destroyer Z-27 was sunk in the Bay of Biscay SW of Ushant along with the torpedo boats T-25 and T-26 on 28 December 1943 by the British cruisers HMS Glasgow and HMS Enterprise. Source: Deutsche Bundesarchive.

At the time we were flying our aircraft from Pembroke Dock to Port Lyautey, six enemy destroyers were known to be at sea, but their location was unknown. While flying through the rain at 3am Christmas Day, 1943, one of our aircraft suddenly encountered flak. The MAD CAT had found the six destroyers. Contact was maintained until the position was checked and rechecked and orders received to proceed to Port Lyautey. This VP-63 contact report resulted in British forces sinking three of the destroyers several hours later.

At Port Lyautey we immediately commenced our operational assignments of convoy coverage and ASW patrols. Soon we learned, however, that U-boats were slipping into the Mediterranean at will. Apparently they were making submerged passages by day at dead slow speed using the current (which is always easterly at shallow depths) to help carry them through the Strait. Ships could

MAD Cats

not stop them because sonar was almost useless due to extreme turbulence in the Strait and the different layers of salinity and temperature. Nets could not be used due to the great depths and turbulence of the water. (A study made by the Office of the Chief of Naval Operations in 1954 stated: "Water conditions were very good for the U-boats.)"

We studied the problem and evolved a plan for a magnetic barrier. The barrier was to be flown as precisely as possible using a conspicuous visual reference point on the shore in Spain and another in Spanish Morocco. The barrier was to be a race track pattern, flown by two aircraft from dawn to dusk at 50 feet altitude with a steep 180 degree turn every minute and twenty seconds. To escape the morning fog at Port Lyautey, we planned for two aircraft to depart Port Lyautey at 11 a.m., fly the barrier from noon to dusk, spend the night at Gibraltar, take off from Gibraltar the next morning, and fly the barrier from dawn to noon at which time they would be relieved by two other MAD Cats coming up from Port Lyautey. There was considerable skepticism regarding our plan and a few senior officers said we were crazy. However, we experienced no difficulty in getting the approval of our bosses and the British. After all, they were desperate. What choice did they have? MAD barrier tactics commenced on February 8, 1944. We continued other ASW patrols, but we gave the MAD barrier top priority over other operational demands.

Several of our experienced plane commanders had been transferred. They had been in action in the far eastern Pacific from December, 1941, until coming to VP-63 in September, 1942, and certainly deserved a rest. However, by this time, our first and second pilots were skilled, experienced, and well qualified to move to plane commanders. We found that flying the barrier was hard work indeed. Also, we were frequently fired upon by Spanish anti-aircraft batteries in Spanish Morocco whenever we were close to their three-mile limit. One of our aircraft was hit by flak, but experienced only minor damages. On February 23, we had a discussion regarding the worth of this effort, but it was quickly decided that we would maintain the barrier unless and until it was definitely proven that U-

The Story of VP-63

boats were slipping through. The following day's events brought the following comments from our civilian scientists, "Yesterday you made a multi-billion dollar decision. This event more than pays for all the money and effort put into the MAD program. Anything further will be icing on the cake." As we would learn later, there would be a great deal of "icing on the cake".

On February 24, 1944, our two MAD Cats on barrier patrol detected, tracked attacked, and sank the U-761. The first magnetic detection occurred at about 3pm. The aircraft proceeded with cloverleaf tracking to "pinpoint" the U-boat for attack. At this time, observing the commotion, the HMS *Wisheart* and the HMS *Anthony* steamed through the pattern of float light markers, scattering them with their wakes. The following voice radio messages took place:

Ship: *"What do you have?"*

Plane: *"A submarine!"*

Ship: *"All you blokes have is a school of fish."*

Plane: *"Get the hell out of here and let someone work on it who knows how."*

The ships then steamed off, but they had caused the MAD Cats to temporarily lose contact. By using a trapping circle, contact was regained at 3:50 and both aircraft attacked at 3:59. Upon hearing explosions of the MAD Cats' contact bombs, the two destroyers returned to the attack position and launched depth charges. Then, a few seconds later, U-761 surfaced and personnel abandoned ship. British destroyers recovered survivors, including the commanding officer. The CO of U-761 was very perplexed. He repeatedly stated that he just couldn't understand how he was hit – he felt certain that there was no ship nearby. We sent our action report to the Chief of Naval Operations in Washington and soon received his reply: "Verify your message. We do not believe this is possible." We felt like adding to our verification, but didn't: "It may not be possible, but we did it just the same."

"Although the German Navy suspected the existence of magnetic airborne detectors, . . . there was no evidence of its use by the Allies:

MAD Cats

the suspicions were aired at a conference held on 10 March 1944 when the head of the development section of the German Naval Technical Signals Departments (*Amtsgruppe Technisches Nachrichtenwesen*), *Kapitan zur See Helmuth Giessler*, informed the assembled officers: "While we employ only acoustic devices (to find submerged submarines) there is a suspicion, which we are examining carefully, that perhaps the enemy is using magnetic-field-change methods as well. The range of such magnetic locators, from ships or from aircraft, is now being examined; from aircraft, a range of between 200 and 400 meters would appear to be theoretically attainable."

"Giessler's statement illustrates, yet again, the paucity of German Intelligence on the Allied anti-submarine measures. By this time, the U.S. Navy had had such a magnetic detection device in service for over a year, and only a fortnight earlier it had played an important part in the destruction of a U-boat."

Detecting U-392 so early in the morning indicated the need to refine our strategy. Something was needed to make the submarines cross our magnetic barrier during daylight hours. We arranged with the British the use of one of their search light aircraft squadrons to flood the approaches to the Strait at night, thus keeping any submarine in the approaches submerged. Having to charge their batteries further to the west would preclude submarines reaching the MAD barrier before daybreak.

Whenever planes could be spared from the MAD barrier, VP-63 continued to fly patrols in the Atlantic and participate in convoy coverage. Not wanting to use the words "MAD Cats" whenever he phoned from Casa Blanca, Commander Moroccan Sea Frontier would say, "Do you have any angry pussies available?"

About this time, the ASW aircraft carrier *USS Block Island* was sunk by a submarine leaving six aircraft in the air with no place to land. Prior to sinking, the ship gave radio instructions to her six aircraft to proceed to the Maderia Islands and ditch their aircraft just offshore. The Casa Blanca command instigated an intensive search centered around the Maderia Islands. VP-63 conducted its own

The Story of VP-63

analysis based on all available facts and factors, and concluded that it was likely that the six aircraft had ditched off the Canary Islands. One of our MAD Cats flew to the Canary Islands and determined that the six aircraft did indeed ditch there. (Of the six pilots from the Block Island, one was rescued the following day by a destroyer and two were returned later by the Spanish authorities.)

Encouraged by sighting a life raft on the water, the VP-63 commander wanted to search as long as possible. To preclude being ordered to return to base too soon, the pilot turned the radio receiver to "off" for several hours. With our technique of leaning the fuel mixture, flying at reduced speed, and flying at 20 feet altitude to secure a cushion effect lift, our endurance was greater than generally thought possible. This flight lasted 24 hours and 5 minutes (with fuel for another hour of flight on board), probably a record for *Catalina* aircraft.

On April 15, 1944, at 3:20 p.m., a magnetic contact was made on the barrier which resulted in the sinking of another submarine. (Months later this was determined to be the U-731.) Again, it was necessary to resort to the aircraft/ship coordinated tactics. The aircraft tracked and then attacked U-731, but U-731 was not hit or not delivered a lethal blow. The aircraft continued to track, and subsequently coached the HMS *Kilmarnock* into attack position. The *Kilmarnock* fired a pattern of hedgehogs at the spot marked by the aircraft. U-731 was hit and went to the bottom.

An interesting sidelight in the sinking of U-731 was the fact that our non-flying Personnel Officer was involved in the action. He had repeatedly stated that a submarine would be caught if we let him go out with a MAD crew – so we finally let him accompany us on April 15. His job was with the K-20 camera during the action

Years later we learned that a Chief of Naval Operations study made in 1954 revealed that two submarines did make successful transits of the Strait. The CNO study stated, "It would appear that the barrier caught three of five of the U-boats. In view of the rough approximations which are involved in establishing the hour when the U-boats entered the barrier area, it is entirely possible that MAD

MAD Cats

barrier activity actually detected every U-boat which passed beneath it.

As weeks passed into months without another magnetic contact on the barrier, we became more and more convinced that U-731 was the last submarine to attempt the passage through the Strait. Months later, we learned that U-731 was indeed the last submarine to attempt the passage and the German Naval High Command had issued orders on May 20, 1944, that "no further U-boats are to be sent to the Mediterranean because losses in the Straits are too heavy." In spite of no contacts, we maintained the magnetic barrier until the submarine was no longer a threat to our forces. We knew that if we took down the "fence", the U-boats would resume slipping into the Mediterranean.

In June 1944 we started returning aircraft, three at a time, to the U.S. for major overhaul. Soon we learned that the Bureau of Aeronautics (BUAER) planned to install bubble bow-turrets on our aircraft. We immediately protested that this would interfere with pilot's visibility and decrease his ability to fly MAD tactics. BUAER replied, "The bubble turrets will be installed because contracts have already been let." We then advised BUAER, "if you install the bubble turrets, we will remove them when our aircraft return to Port Lyautey." The bubble turrets were not installed.

After April 15, the number of U-boats in the Mediterranean steadily declined, since losses were not replaced. By August, the few U-boats remaining in the Med were bottled up in their ports. No operational U-boats were left – not one. At this point in time, British Admiral Cunningham sent us the following message, "Your MAD Cats have turned the Mediterranean into an Allied lake." Vice Admiral Hewitt verbally told us the same thing when he visited Port Lyautey in July 1944. Later a Navy Department press release dated July 7, 1945, stated, "In the process of turning the Mediterranean into an Allied lake, Patrol Bombing Squadron Sixty-Three and its Gibraltar U-boat 'fence' is credited with being a major factor in the successful invasion of Southern France, since no Allied vessel was sunk by an enemy submarine." All first contacts made

The Story of VP-63

on U-boats in or near the Strait had been made by MAD Cat aircraft. Three U-boats had been sunk in the Strait – with two actions assisted by British ships. In effect, the gate was "closed and locked" by airborne magnetometers, millions of tons of shipping in the Mediterranean was now safe, and the successful invasion of Southern France was assured. The Navy Department press release dated July 7, 1945, stated, "This squadron is credited officially with sinking three under seas craft and is listed as having inflicted probable damage to others." Clarification of the words "probable damage to others" follows.

On April 20, 1945, a VP-63 pilot on patrol sighted a snorkeling submarine. He dove from 2,000 feet, flew up the wake over the still visible snorkel, and on the first peak swing of the magnetic signature, fired his pattern of 30 retro contact bombs. Evidence of explosions and evidence on the surface, including a 2,700 yard long oil slick, was very convincing evidence that this submarine had been sunk. Also, no mention is made in the Navy Department Press Release of the submarine that surrendered to a VP-63 aircraft about one month before VE Day. A VP-63 aircraft on patrol sighted a fully surfaced submarine. As the pilot closed in for the attack, the submarine raised a large white flag as a sign of surrender. The aircraft maintained close contact until a British destroyer arrived to accept the surrender and take the submarine in tow. At this point in time, VP-63 should have been credited with sinking four submarines and capturing a fifth.

U-541 Surrenders to MAD Cat 11 May 1945 Source: VP-63

MAD Cats

The words "probable damage to others" no doubt also refers to about 15 other attacks which may or may not have been on submarines. Squadron policy was to track and attack any magnetic anomaly which have a characteristic submarine signature. The anomaly could be caused by a submarine lying on the bottom rather than an uncharted wreck. Moreover, we had to expend our torpex bombs before they became over age and dangerous. It is likely that at least one of these attacks was on a submarine and the submarine was destroyed – especially the attack near a periscope sighting by an RAF aircraft made about one hour prior to our attack.

A detachment of the squadron returned to the United Kingdom during the closing months of the war to participate in the attacks that marked the final flurry by the U-boats.

On July 7, 1945, Patrol Squadron Sixty-Three, the first U.S. Naval Aviation squadron to operate from the United Kingdom and the only squadron to be loaned to the RAF in the war against German submarines, returned to the United States.

After specialized training by Air Force, Atlantic Fleet, personnel of VP-63 were transferred to the Pacific Theater of War.

The story of Patrol Squadron Sixty-Three is not complete without a few words regarding the character of our personnel. There was not a disciplinary action needed which required a Captain's Mast, no one was ever over leave or over liberty, there was no grumbling over the dangers, bedbugs at Gibraltar, poor food in the UK, etc. Our personnel worked long and hard, far beyond what was asked of them, to keep the planes flying and to keep the equipment at peak efficiency. The moral character of our personnel was probably a cut or two better than the average military unit, but there is another reason. We were driven by some unseen force, perhaps God Himself; we all felt it. We were "fired up". We took some time for needed rest, exercise, and recreation; but we didn't have time for nonsense.

Due to the secrecy of our operations and the fact that few would believe our "impossible" accomplishments anyway, VP-63 never

The Story of VP-63

did receive the credit it so richly deserved – either as a unit or as individual members of the squadron. Many of our personnel were "short-changed" in regard to medals and subsequent promotions, which usually accrue following truly outstanding performances of duty. It would have been satisfying and rewarding to have had appropriate recognition. However, medals and promotions are not too important. We have a reward far greater than medals and promotions, a reward that will always remain with us - we dared to do the impossible in response to the will of God and made a truly significant and far reaching contribution to our nation and the war effort. This fact gives us great joy, peace, and inner satisfaction. That is our reward.

July 1945 marked the official termination of Patrol Squadron Sixty-Three, but the story of the MAD Cats was far from over. Patrol Squadron Sixty-Three personnel were scattered throughout the Navy and were destined over the years to exert a profound influence on weapon systems and tactics for airborne anti-submarine warfare. This influence was expressed in 1954 by Admiral Baker, Commander Operational Development Force, when he stated, "Patrol Squadron Sixty-three is the father of modern airborne anti-submarine warfare." Changes, however, were very slow in coming.

The reluctance of the Navy to adopt new ideas and concepts of proven value is indicated by the words of the Chief of Naval Operational Analysis in the Navy Department when he stated in a speech in Washington, DC in 1946, "I would recommend MAD for ASW aircraft if the next war were to be fought in the Strait of Gibraltar; otherwise, no."

Shortly after VX-1 Anti-Submarine Development Squadron was commissioned, they evaluated MAD and concluded that the system had very limited value. What they failed to recognize was that successful detection from a trapping circle or expanding spiral search, followed by tracking and "pinpointing" for attack, required a high degree of skill on the part of the pilot and MAD operator, and this skill could be obtained only with practice. Then, in 1950, a few pilots and technicians from VP-63 were assigned to VX-1. It did not

MAD Cats

take long to demonstrate and convince all personnel in VX-1 that MAD is effective when proper tactics are employed by skilled personnel. Convincing the rest of the Navy was not so easy.

In 1952 a VP-63 officer was ordered to command VX-1, and an intensive program was begun to indoctrinate senior naval officers by letting them observe from the co-pilots seat. By this time our "hot shot" MAD pilots were flying at 20 feet altitude, pulling up to 50 feet in the turns, and maintaining contact against fast and maneuvering submerged "tame" submarines. Admiral after Admiral would step out of the aircraft after demonstrations, visibly shaken by the low altitude tactics, but convinced of the effectiveness of MAD. By 1954, the Navy was convinced and action was taken to install the MAD system on all Navy ASW aircraft, including carrier-based aircraft. It is interesting to note that in about 20,000 hours flown at extremely low altitude, there had not been a single aircraft lost due to this low-level flying.

There is more to this story than magnetic anomaly detection. Patrol Squadron Sixty-Three was the first squadron to employ sonobouys and other equipment and to develop tactics for their use. Personnel from VP-63, which included the commanding officers of VS-30 and VX-1, conceived of new weapon systems and made improvements on existing weapons systems. For example, personnel from VP-63, while attached to VX-1, originated the *Project Julie* system and they and other VX-1 personnel proved its merit.

Patrol Squadron Sixty-Three and the science and technology out of which it was born live on in today's sophisticated anti-submarine aircraft of the U.S. Navy.

The original MAD Cat squadron lives on – in the MAD and other ASW equipment installed in modern ASW aircraft, in the official U.S. Navy tactical publications which specify the tactics which were developed by VP-63, and in the minds and hearts of men who dare to answer God's call to do the impossible.

> When the roll is called up yonder,
> In the land of sweet bye and bye,

The Story of VP-63

They shall have a great reunion
In their home up in the sky.

PBY-5 At Pembroke Dock, UK. Source: U.S. Navy

Pembroke, Wales in 1944 (Source VP-63)

In *Wagner's Men*, Captain Curtis H. Hutchins added the following:

Errors in the history books:

P. Creme, *U-Boat Commander*, 1984 Naval Institute Press, 1984, lists the following as being responsible for sinking U-761 on 26 Feb 1944 – VP-63, VB-127, British Squadron 202, HMS *Wishart*, and HMS *Anthony*. We all know that the VB-127 aircraft and the British Squadron 202 aircraft made their attacks on a helpless and sinking submarine after the U-Boat crew had abandoned ship. The depth charges from these attacks killed some of the U-boat crew, thus losing some valuable intelligence information.

MAD Cats

The *Anthony* and *Wishart* interfered with the MAD tracking pattern and almost enabled the U-Boat to escape, and did cause contact to be lost for about fifty minutes. While the depth charge attacks of *Anthony* and *Wishart* inflicted additional damage to the U-boat, there is no doubt that the VP-63 aircraft struck the lethal blow. The surviving U-boat commander stated several times to our interrogating intelligence officers that "the FIRST set of explosions caused such severe damage that it was a case of surfacing now or never, and the second an third set of explosions added to the damage." Intelligence officers said VP-63 deserves ALL the credit for sinking U-761. The other units participating in the action did more harm than good.

Other comments:

From the beginning, we had a cooperative attitude. Eddie Wagner provided the spark we needed to "fire us up." We realized that the motto he gave us was really true – that there are few things in this life that are truly impossible if we act as if it were impossible to fail.

Motto: "The difficult we do immediately...the impossible takes a little longer."

Many little non-glamorous acts led to big success.

One night at Port Lyautey the MAD Cat CO went down to the hanger at 1am and found a maintenance crew working on an aircraft. The C.O. said, "This aircraft is not scheduled to fly; why are you working on it at this hour?" The chief in charge replied, "Skipper, something might happen – you might need this aircraft." He was right. Something did come up and we did need the aircraft.

There is another reason for our success. We were not only sharper than other squadrons, we were also smarter than the commands over us. To cite a few examples:

First: the wing at Bermuda wanted us to look for a downed *Coronado* aircraft to the southwest. We flew to the northeast and found the survivors in two life rafts.

The Story of VP-63

Second: the wing at Port Lyautey wanted us to look for downed aircraft from the Black Island in the vicinity of the Madeira Islands. We flew to the Canary Islands and found the aircraft ditched there.

Third: the wing at Port Lyautey was very skeptical of our prediction of the time a U-Boat would cross our MAD barrier and of our sending a third aircraft to Gibraltar to provide additional coverage for the morning. U-731 was detected within 30 minutes of the predicted time.

Fourth: Commander, Moroccan Sea Frontier, argued that if two aircraft provided a fifty percent probability of detection on the MAD barrier, than four aircraft would provide one hundred percent probability of detection. With considerable effort the MAD Cat CO was able to convince them that four aircraft would raise the probability from fifty percent to seventy-five percent.

One other reason for our success was the encouragement of our loved ones back home.

Historians tell us that history does not become reasonably accurate until at least fifty years after the event. So, we have a few more years to go before the general public has an accurate view of the history of VP-63.

Capt Curtis H. Hutchins

A Special Assignment

When the squadron transferred to Port Lyautey, French Morocco around 13 December 1943, two planes, No. 4 with Lt V.M. Mayabb as Patrol Plane Commander, and No. 15, with Lt R.A. Barton as Patrol Plane Commander, remained in England for special work with the Royal Air Force. Mr. E.W. Westrick, the civilian MAD technician attached to the squadron, also remained for this work. As the only MAD-equipped planes in operation in the United Kingdom, they were (supposedly) needed for urgent experimental work in connection with counter-measures being developed by the British authorities against the flying bomb (V-1), which they knew the

MAD Cats

Germans were then preparing. These two planes completed their work, which included flights over the London area, and rejoined the squadron at Port Lyautey on 20 January 1944.

While there was no other clarification as to the mission, discussions with Stuart Slade, Senior Naval Analyst for Forecast International on August 19, 1993, came up with the following:

In 1943, the British had started experimenting with midget submarines. They knew the Germans were developing 4-, 6-, and 8-man midget subs; and most of British harbors are at the end of long estuaries, which are very vulnerable to midgets. In fact, in 1945 the Germans Type XXIII did penetrate quite deep into British coastal waters. Slade thinks that the MAD Cats were held back for experiments to see if MAD could detect midgets in estuaries and other such shallow water. "That is about the only thing I can think of fitting the time line. I couldn't have been anything hairy because the *Catalina* was already considered too vulnerable to send out into opposed airspace, so it had to be used close in to shore. It may have been something just as simple as basic trials. The MoD was getting very interested in midgets about that time, and were expecting attacks by midgets."

A German Molch class midget submarine. In the last years of WW2 the Germans built large numbers of such small submarines. Source: Deutsche Bunderarchive

The true mission of the deployed aircraft may never be fully explained. Another possibility could have been an involvement in one of the most secret operations of the war. In his book *Spec Ops – Case Studies in Special Operations,* Admiral William H. McRaven, USN, the current Commander of United States Special Forces, discusses the Allies effort to develop a midget submarine which

The Story of VP-63

could attack the German battleship *Tirpitz* sister ship to the *Bismark*. The goal was to prevent the dreadnought from becoming a threat in the North Sea (*Spec Ops*, p.201).

British midget submarines carried out a damaging attack on the Tirpitz. Source: U.S. Navy

While the author could find no records which specifically point to VP-63 tasking, the time frame would fit, either as a test of the ability of MAD to detect one of the X-craft during trials, or possibly to search for one of the tricky little boats that sank during testing.

The MAD Cats do not say what the special deployment was for. Given the extreme secrecy of the effort (*Spec Ops*, p. 210), the crews may not have known what they were looking for.

MAD Cats

The Tall Tales
In their own words

While World War II did a lot for technology, it also brought out the best in our fighting force: courage, creativity, and a unique way of looking at a bad situation. This is something that makes the warrior, especially the American warrior, unique. It is something VP-63 had in abundance.

As World War II becomes more distant, the number of veterans of that war are disappearing. In 2012 the Veterans Administration estimated that 1.7 million were still alive; but they were passing away at the rate of nearly 250,000 a year; with the rate increasing as the vets age. The concern that the world would lose the stories of those who lived the experience, a national treasure, prompted Congress to establish the Veterans History Project in 2000. The Library of Congress has an active, ongoing program to capture the stories of veterans while it can, but it is feared that only one in two hundred stories will be preserved. Other, private efforts are encouraged. This is one.

MAD Cats

Squadron Rosters
VP-63

Officers – 1943

CMDR. C. H. Hutchings, USNA '34
LT. CMDR. C. W. Brown, A-(N)

Lieutenant Virgil M. Mayabb, USN
Lieutenant Robert A. Barton, USNR
Lieutenant Lyman C. Beardsley, USNR
Lieutenant Carl E. Benscoter, USNR
Lieutenant, James W. Hardy, U.S.N.
Lieutenant Norman S. Johnson, USNR
Lieutenant Marshall W. Nicholson, USNR
Lieutenant Woodrow E. Sholes, USNR
Lieut.(jg) William S. Andrus, USNR
Lieut.(jg) Robert I. Bedell, USNR
Lieut.(jg) Cornelius D. Brislawn, USNR
Lieut.(jg) Roderick M. Brush, USNR
Lieut.(jg) Stanley H. Castleton, USNR
Lieut.(jg) James W. Christ, USN
Lieut.(jg) Marvin S. Clinton, USNR
Lieut.(jg) Luis F. Corea, USNR
Lieut.(jg) William P. Craddock, USNR
Lieut.(jg) Robert H. Desbrow, Jr., USNR
Lieut.(jg) John R. Dutcher, USNR
Lieut.(jg) John S. Elliott, USNR
Lieut.(jg) Rowland M. Fairlie, USNR
Lieut.(jg) John C. Fox, USNR
Lieut.(jg) Francis C. L. Head, USNR
Lieut.(jg) Edward M. Hillie, USNR
Lieut.(jg) Edwin W. Kellogg, USNR
Lieut.(jg) Gilbert L. Knight, USNR
Lieut.(jg) Rex B. Knorr, USNR
Lieut.(jg) Van A. T. Lingle, USNR
Lieut.(jg) Billy May, USNR
Lieut.(jg) Walter J. McHargue, USNR
Lieut.(jg) Charles A. Merryman, USNR

The Story of VP-63

Lieut.(jg) Elmer D. Moone, USNR
Lieut.(jg) Joseph A. Pariseau, USNR
Lieut.(jg) Sam R. Parker, USNR
Lieut.(jg) Clinton R. Rehn, USNR
Lieut.(jg) Melbourne "J" Simer, USNR
Lieut.(jg) Ralph C. Spears, USNR
Lieut.(jg) Matthias J. Vopatek, Jr., U.S.N.
Lieut.(jg) Thomas R. Woolley, USNR
Lieut.(jg) Hubert L. Worrell, USNR
Ensign Howard J. Baker, USN
Ensign Lyle E. Bonn, USNR
Ensign Gaddis G. McKee, USN
Gunner Earl C. Bonner, USN
Gunner Joseph A. Fahrner, USNR
Machinist Harry M. Tyler, USN
Civ. Tech. Edmond W. Westrick.

Chiefs -- 1943

Loren A. Alford, ACMM(PA), USNA
Michael Babey, ACRM(AA), USNA
Edward F. Bradshaw, Jr., ACMM(AA), USNA
Ehler F. Clausen, ACMM(AA), USNA
William E. Forland, ACRM(AA), USNA
Leo P. Harris, ACRM(AA), USNA
Norman A. Hoy, CBM(PA), USNA
Stanley E. Loughlin, ACMM(AA), USNA
Roderick S. McCrae, CAP(AA), USNA
John W. McKeown, ACOM(AA), USNA
William F. McSharry, CAP(AA), USNA
Anthony Mule, CAP(AA), USNA
Wynne T. Ramsey, ACRM(AA), USNA
Edward Rex, ACM(PA), USNA
Vivian L. Robison, ACMM(PA), USNA
Max W. Rood, ACMM(AA), USNA
Frederick L. Rosenmund, ACMM(AA), USNA
Olin R. Ruff, CAP(AA), USNA
James W. Skelton, ACMM(AA), USNA
Vito J. Soranno, CAP(AA), USNA
Robert J. Watson, Jr., CY(AA), USNA
Herbert J. Yeck, ACMM(A4), USNA

MAD Cats

Enlisted 1943

Clarence S. Abbott, AMM3c, USNA
George D. Ackerman, Jr., OM2c, SNA
Hedley T. Adams, AMM2c, USNA
Bernard L. Alexander, AOM3c, SNA
Floyd L. Baas, ARM2c, USNA
Harry N. Beam, ARM2c, USNA
Theodore M. Benedict, Jr., AMM2c, USNA
William R. Berge, Jr., AMM3c, USNA
Wilson E. Black, ARM2c, USNA
Harry M. Bolsinger, A0M2C, USNA
Placido A. Bonanno, ARM3c, USNA
Charles A. Borland, AOM3c, USNA
Frank B. Bowen, AP1c, USNA\
Guy D. Bowers, AMM2c, USNA
Walter E. Brown, ARM1c, USNA
Lee D. Bulloch, AMM2c, USNA
Olin F. Bunch, ARM1c, USNA
James J. Byrd, ART1c, USNA
Robert L. Cantrell, AMM2c, USNA
Anthony J. Catino, Jr., AOM3c, USNA
Adam J. Chaisson, ARM3c, USNA
Joseph P. Chiapulis, AOM1c, USNA
Edwin P. Christenson, AMM3c, USNA
Lester M. Coker, AOM3c, USNA
Thomas A. Comstock, AOM3c, USNA
Milton H. Connery; AOM3c, USNR.
James A. Correll, AOM1c, USN.
George W. Crane, Jr., AOM3c, USNR.
Marlin Crider, ARM1c, USNR.
Melvin B. Cummins, AMM1c, USN.
Carl E. Danielson, ART1c, USNR.
Leslie O. Densley, AMM1c, USNR.
Max Dolloff, AMM1c, USN.
Lloyd S. Donley, AMM1c, USN.
Clifford J. Donth, AMM3c, USNR.
Joseph F. Douglass, AOM3c, USNR.
Erick A. Engman, AMM2c, USNR.
James J. Evans, AMM1c, USN.
James C. Ferris, Jr., AP1c, USN.
John C. Fisher, ART1c, USNR.
William N. Franklin, AMM2c, USNR.
James R. Fritz, AMM2c, USN.
Fred Futterman, AMM3c, USNR.
Carl E. Gaskin, Jr., AMM2c, USNR.
Wallace N. Gooch, ARM3c, USNR.
Charles L. Gravel, ARM3c, USNR.
Kenneth E. Green, AMM3c, USN.

Gilbert A. Hall, ART1c, USNR.
Charles J. Hargrove, ARM2c,
Charles H. Haupt, Jr., ARM2c, USN.
William A. Hawksby, Y3c, USNR.
William W. Heath, AP1c, USN.
Bobby G. Henderson, ARM3c, USN.
James N. Henson, ARM1c, USN.
Lawrence J. Hickey, Jr., AMM1c, USN.
Selix H. Hill, AMTc, USN.
Delbert D. Hubbartt, ARM2c, USN.
Jack E. Hughes, ARM2c, USNR.
Walter H. Irby, ART1c, USNR.
Joseph V. Jellison, AOM2c, USNR.
Charles W. Jones, S2c, USNR.
Silas Katz, S2c, USNR.
Milton F. Keane, AOM3c, USNR.
John J: Ketterman, AP1c, USN.
Charles C. King, AMM2c, USNR.
Dean E. Kinkel, ART1c, USN.
John K. Kirkpatrick, AOM, USN.
James A. Kohl, Jr., AOM, USNR.
Amos C. LaMora, AOM3c, USN.
Bernard N. Lawson, AMM2c, USN.
George T. Ledoux, AOM3c, USNR.
Howard G. Lee, AP1c, USN.
Richard L. Lewis, AOM2c, USN.
John D. Linhart, AMM3c, USNR.
Paul H. Lyons, AOM3c, USNR.
James F. Madden, S2c, USNR.
Salvatore P. Maneen, S1c, USNR.
James O. Manning, AMM2c, USNR
Billy F. Martin, AMM2c, USN.
John S. Mason, ARM3c, USNR.
Floyd L. Matthews, PhM1c, USN.
John V. Mauch, ARM2c, USNR.
Delbert I. Mayhew, AMM2c, USN.
Bernard E. Mazza, AOM2c, USNR.
Willie V. McCormick, AMM3c, USN.
Miles E. McKenna, Jr., ARM2c, USNR.
Michael H. McLane, Sec, USNR.
Herbert J. Miltz, Jr., ARMic, USN.
John H. Olsen, ARM2c, USN.
James E. Peaden, ART1c, USNR.
Francis R. Pearson, AMM2c, USN.
Walter I. Perry, AP1c, USN.
Richard G. Peterson, ARM1c, USNR.

The Story of VP-63

Benjamin F. Pfannstiel, AMM2c, USN.
Marion C. Repass, AMM2c, USN.
Joe G. Reyes, S2c, USNR.
Owen R. Rice; ARM2c, USNR.
Loring R. Richardson, AP1c, USN.
Norman L. Rogers, ARM1c, USN.
Joseph A. Scarsella, AMM2c, USNR.
Arthur J. Schneider, AMM1c, USN.
Max L. Shaw, S2c, USNR.
George B. Sipe, ARM2c, USNR.
Burton D. Smith, AP1c, USN.
Bruce Smithee, AP1c, USN.
Anthony V. Spell, AMM2c, USN.
William D. Spence, Jr., AMM1c, USN.
Albert W. Stevenson, SK1c, USNR.
Harvey J. Sutton, SK1c, USNR.
George E. Swink, ARM2c, USNR.
Earl H. Tanneberg, AMM2c, USNR.
Floyd I. Tanner, AMM3c, USNR.
Frank J. Tanner, AMM3c, USNR.

Wethrel W. Taylor, S2c, USNR.
Thaddeus T. Tull, ARM1c, USNR.
Paul J. Turner, AMM3c, USNR.
George L. Tiller, Sec, USNR.
John W. Vertes, AMM3c, USNR.
Joseph M. Walker, ARM3c, USNR.
Paul F. Ware, AMM1c, USN.
Roy E. Warner, Jr., S2c, USNR.
Wilhelm H. Watermann, ARM3c, USNR.
Raymond I. Weinrub, ARM2c, USN.
Robert H. Whalen, AP1c, USN.
Franklin C. White, Jr., ARM2c, USN.
Reuben S. Wilkins, AMM3c, USNR.
James A. Willey, AOM2c, USNR.
Edward G. Woods, AMM1c, USN.
Joseph Yourek, Sec, USN.
Joseph Zelenka, Jr., AOM2c, USNR.
Harry Zukerman, AMM2c, USNR

MAD Cats

Officers 1944

CMDR. C. H. Hutchings, USNA '34
LT. CMDR. C. W. Brown, A-V(N)

LT. M. J. Firey, A-V(S)
LT L.D. Russell, A-V(G)
LT J.W. Hardy, USN
LT. M.W. Nicholson, A-V(N)
LT C.A. Benscoter, A-V(N)
LT. R.A. Childers, A-V(N)
LT N. S. Johnson, A-V(S)
LT F. G. Lake, A-V(G)
LT. R.A. Barton, A-V(N)
LT R. C. Spears, A-V(N)
LT S. R. Parker, A-V(N)
LT J. C. Brown. Jr., A-V(N)
LT William D. Ray
LT H. L. Worrell
LT John M. Nester
LT L. C. Destowlinski
LT H. J. Baker
LT B. May
LT S. H. Castleton
LT R. H. Desbrow
LT T. R. Woolley
LT J. A. Pariseau
LT R. B. Knorr
LT E. D. Moone
LT M. J. Simer
LT J. R. Dutcher
LT F. C. L. Head

LT (jg) J. E. Elliott, A-V(N)
LT (jg) W. P. Craddock, A-V(N
LT (jg) C. Fox, A-V(N)
LT (jg) W. S. Andrus, A-V(N)
LT (jg) J.I. Munson, A-V(S)
LT (jg) Donald F. Young, A-V(N)
LT (jg) Russell E. Asper, A-V(N)
LT (jg) Robert P. Hatter, A-V(N)
LT (jg) McClure, A-V(N)
LT (jg) J.W. Christ, USN
LT (jg) M.J. Vopatek, USN
LT (jg) R.C. Williams, USN

The Story of VP-63

LT (jg) R.M. Fairlie, A-V(N)
LT (jg) V.A. T. Lingle, A-V(N)
LT (jg) C.A. Merryman, A-V(N)
LT (jg) W. J. McHargue, A-V(N)
LT (jg) E.M. Hillie, A-V(N)
LT (jg) R.I Bedell, A-V(N)
LT (jg) G.L. Knight, A-V(N)
LT (jg) E.W. Kellogg, A-V(N)
LT (jg) C.R. Rehn, A-V(N)

ENS. W G. Hennings, A-V(N)
ENS Jack J. Pritchard, A-V(N)
ENS. W.H. Ritter, A-V(N)
ENS. G.G. McKee, A-V(N)
ENS. J.E. McDougall, A-V(N)
ENS. E.F. Horne, A-V(N)
ENS. B.C. Wyrick, A-V(N)
ENS E.W. Hill, A-V(N)
ENS. W.W.H. Ross, A-V(N)
ENS. E.T. Koob, A-V(N)
ENS. R.D.J. McCarty, A-V(N)
ENS. H.O Gillespie, A-V(N)
ENS. K.C. Peterson, A-V(N)
ENS. E.W. Zinger, A-V(N)
ENS. W.F. McSharry, USN
ENS. O.R. Ruff, USN
ENS. W.I. Perry, USN
ENS. Bruce Smithee, USN
ENS. H.I. Adelt, A-V(N)
ENS. J.C. Logue, A-V(N)
ENS. W.I. O'Brien, A-V(G)
ENS. Tadeuse Macur, A-V(N)
ENS. J. L. McShane, A-V(N)
ENs. N.B. Primm,. Jr., A-V(N)
ENS. F.B. Bowen, USN
ENS. L.R. Richardson, USN
ENS. A.M. Williams, A-V(S)

Gun. E.C. Bonner, USN
Gun. J.A. Fahrner, A-V(S)
ACC; Edmond W. Westrick, US Tech.
Ch. Mach. J.J. Menzaco, USN

MAD Cats

CHIEFS 1944

Leading Chief Frederick L. Rosenmund ACRM
Michael Babey ACMM
Edward F. Bradshaw, Jr. ACMM
Ehler F. Clausen ACRT
John C. Fisher ACRT
Gilbert Ashley Hall ACRM
Leo P. Harris ACMM
Lawrence J. Hickey, Jr. ACRT
Harold Irby CAP
Howard G. Lee ACMM
Stanley E. Loughlin ACM
Hugh McCoy CAP
Roderick S. McCrae ACOM
John W. McKeown CAP
Anthony Mule ACMM
Michael Post ACMM
Arthur L. Pruitt ACRM
Wynne T. Ramsey ACMM
Vivian L. Robison CAP
Burton D. Smith CAP
Vito J. Soranno ACMM
William D. Spence, Jr. ACMM

The Story of VP-63

Enlisted Roaster – 1944

Clarence S. Abbott AMM
George D. Ackerman, Jr. AOM
Hedley T. Adams AMM
Bernard L. Alexander AOM
Floyd L. Baas ARM
Clarence J. Barry AMM
A. L. Bell AOM
Harry N. Beam ARM
Theodore M. Benedict AMM
William Richard Berge AMM
Wilson E. Black ARM
Harry M. Bolsinger AOM
Placido A. Bonanno ARM
Charles A. Borland AOM
Guy D. Bowers AMM
Clarence J. Brown AMM
Leo D. Bulloch AMM
Olin F. Bunch ARM
William A. Butts ARM
Robert L. Cantrell AMM
H. E. Carling ARM
John W. Carthel, Jr. S2C
Anthony J. Latino, Jr. AOM
Adam J. Chaisson ARM
Murray Charnin ARM
Joseph P. Chiapulis AOM
Lester M. Coker AOM
Thomas J. Collier ARM
Ernest H. Colomb ARM
Thomas A. Comstock AOM
Milton H. Connery AOM
Harry J. Cooper ARM
James W. Cooper AMM
James A. Correll AOM
James P. Cosgrove ARM
George W. Crane, Jr. AOM
Marvin Crider ARM
Melvin B. Cummins AMM
James A. Cunningham ARM
Carl Emil Danielson ART
Alexander Dervech AMM
Max Dolloff AMM
Lloyd S. Donley AMM
Clifford J. Donth AMM
Joseph F. Douglass AOM

Erick A. Engman AMM
James Jeremiah Evans AMM
William N. Franklin AMM
James R. Fritz AMM
Fred Futterman AMM
Carl E. Gaskin, Jr. AMM
Wallace N. Gooch ARM
Charles L. Gravel ARM
Kenneth E. Green AMM
John A. Griffin. AMM
Charles M.W. Hanson ARM
Charles J. Hargrove ARM
Charles Haupt ARM
William A. Hawksby Yeoman
William W. Heath AP
Bobby G. Henderson ARM
James N. Henson .ARM
Francis S. Higgins AMM
Selix Hunter Hill AM
H.C. Hollister ARM
George T. Holmes ART
Delbert D. Hubbartt ARM
Jack E. Hughes ARM
Walter L. Hughey S2C
C.V. Hunt ARM
Joseph V. Jellison AOM
Carl E. Jolly, Jr. ARM
Charles W. Jones AM
Silas Katz AMM
Milton F. Keane AOM
Marion H. Kent AMM
Charles C. King AMM
Dean E. Kinkel ART
John K. Kirkpatrick AOM
James A. Kohl, Jr. AOM
M.A. Kolanda AMM
E.D. Kravick ARM
Amos C. La Mora AOM
Bernard N. Lawson AMM
George T. Le Doux AOM
Richard L. Lewis AOM
Robert F. Light ARM
John D. Linhart AMM
Paul H. Lyons AOM
James F. Madden SKV
Salvatore P. Maneen . Yeoman
Billy F. Martin AMM

MAD Cats

John S. Mason ARM
John V. Mauch ARM
Delbert I. Mayhew AMM
Bernard E. Mazza AOM
Charles J. McBride ARM
George V. McBryde ARM
T.L. McClure ARM
Willie V. McCormick AMM
H.M. McGuire, Jr. AOM
Miles E. McKenna, Jr. ARM
Michael H. McLane AMM
Charles W. Miller AMM
Herbert J. Miltz, Jr. ARM
Colonel J. Moore, Jr. AMM
Alexander J. Mundell AMM
Walter H. Murren AMM
John H. Olsen ARM
Vincent C. Paradiso S2C
Douglass C. Peterson AOM
J. Pawlak AMM
James E. Deaden ART
Francis R. Pearson AMM
Richard G. Peterson ARM
Alphonse N. Petrocci ART
Benjamin F. Pfannstiel AMM
C.W. Platte AMM
P.G. Rawlings ARM
Marion C. Repass AMM
Owen R. Rice ARM
E.C. Richter ARM
William G. Riggins AMM
John D. Robertson, Jr. AMM
Norman L. Rogers ARM
G. C. Ross AMM
Donald W. Rusbuldt AMM
H.H. Schrant AMM
Joseph A. Scarcella AOM
Charles K. Schroeder AMM
William R. Selman AMM
Stanley L. Selzer ART
Max L. Shaw ARM
Robert G. Shirley AMM
James A. Shultz AMM
George B. Sipe ARM
Michael F. Socha AMM
Lawrence E. Sparks AMM
Anthony V. Spell AMM
William J. Spencer ARM
Albert W. Stevenson SKV
George W. Stokes AEM

Doyce R. Stroup ARM
Harvey J. Sutton SKV
Myron C. Tallman AMM
Earl H. Tanneberg AMM
Floyd I. Tanner AMM
Frank J. Tanner AMM
R.A. Taylor AMM
Wethrel W. Taylor AOM
Thaddeus T. Tull ARM
Paul J. Turner AMM
John W. Vertes AMM
Joseph M. Walker ARM
Paul F. Ware AMM
Roy E. Warner, Jr. AOM
Arthur E. Weeks AMM
Raymond I. Weinrub ARM
R. W. Whelan ARM
Lawrence R. Whiffen AMM
Franklin C. White, Jr. ARM
Ira C. Wicker AMM
P.H. Wireman ARM
Reuben S. Wilkins AMM
Edward J. Willenborg ARM
James A. Willey AOM
Edward G. Woods AMM
Joseph Yourek AM
W.L. Zaida ARM
Joseph Zelenka AOM
Harry Zuckerman AMM
Lloyd K. Anderson AMM
Leonard Annable, Jr. AEM
David Aronin AMM
Ferdinand P. Actman AOM
Samuel Azrael AMM
Thomas J. Babine, Jr. ARM
Richard D. Ballew, Jr. AMM
William W. Bard AMM
Billy Beam Slc
Donald W. Beaty AMM
Gordon H. Beckman AMM
Robert W. Bell AMM
Lloyd Bell S2c
Richard M. Bernard S1c
John J. Black AMM
Howard T. Boland AMM,
Cecil H. Bowen S1c
James F. Boyd AM
D.L. Brice S2c
Wayne G. Brinson AOM
Charles E. Britton S1c

The Story of VP-63

W.R. Brosious S2c
William H. Burns AMM
Malley J. Byrd AMM
Dominic Champagnone S2c
Nicholas C Arnaggio AMM
Ralph W. Chapdelaine AMM
Harry A. Chojnowski AEM
R.W. Christenson S1c
A A. Clarke S1c
B.J. Clayton S2c
Odie W. Coffey AMM
Charles J. Consiglio AMM
C. Costin S2c
Henry T. Coy AEM
Hale L. Cummings, Jr. AMM
Henry C. Daniels AMM
Wilfred H. Daringer AMM
Stanley J. Dash AMM
Ariel B. Davis AMM
Robert J. Davis AMM
Leslie T. Dent S2c
Robert W. Derrer AMM
Clifford Dixon S1c
J.L. Dodd S2c
D.W. Eaddy S2c
James A. Ellis AMM
C.M. Elmore S2c
Clarence E. Emmons S1c
Charles E. Essman AMM
James Evans, Jr. AMM
W. L. Ewing S2c
Dominic Ferraguto AOM
John Ralph Finkbone AMM
Ercil J. Foster AM
James R. Gold AOM
Lewis A. Greene AMM
Charles A. Greene AMM
Jesse N. Gross AMM
Herbert E. Groves AMM
Anthony Gualtieri AOM
Joseph E. Gulledge S1c
Francis M. Haley AMM
Robert B. Harrington AMM
Burton P. Harris ARM
Melvin Edward Harris AMM
Clifford G. Hawkes ART
Robert W. Hawley AMM
Charlie E. Head S2c
Hershel W. HIGGS S1c
Jonathan J. Hultz AMM

Earl W. Hunt AOM
Edward J. Janenka AOM
Edward E. Jarsabski AOM
Charles R. Johnson AMM
Karl C. Jones AMM
William H. Jones AMM
Frank J. Kline, Jr. AMM
Leonard Kogon AMM
Clifford H. Lamprey AOM
Jean D. Lablanc AOM
Glenn E. Legacy S2c
Richard A. Lessard S2c
Alva Lewallen AMM
Robert G. Lodge S2c
Frank R. Loeffler AEM
Carl J. Long AMM
John Lovas S2c
Albert Lunn AMM
Saul Lustberg S2c
Duncan A. MacLeod AMM
Bertrand H. Mahoney ART
Lewis Malinowski AMM
Robert J. Mason AMM
Robert M. McDonough AMM
Guy B. McNeil. ART
Jack H. Mermod AMM
Raymond J. Mikus AMM
Charles J. Montefusco AMM
Morris H. Moses AMM
William M. Nabers AMM
Matthew Ogrin ARM
Blake H. Page AMM
Lawrence P. Perkins S2c
Warren H. Peterson AEM
Peter P. Piaskowski S2c
Colin C. Primrose AMM
Francis F. Protho S1c
Vincent J. Reilly AEM
Thomas N. Renfro S2c
Joe G. Reyes AMM
Howard C. Rich, Jr. AMM
James S. Riley AM
Irving Rogers AOM
Kenneth R. Root AM
Joseph W. Ross AOM
Rudolph R. Russell AOM
Walter F. Shanahan ARM
Thomas W. Sharratt AOM
Clair Shirley, Jr. S2c
Albert C. Sinkus AMM

MAD Cats

Daniel Thomas Slattery AEM
W.E. Snodgrass S2c
Carl E. Snyder AMM
Samuel L. Spinks, Jr. AMM
Sam Siam AMM
Ray S. Stein ART
William V. Stevenson AMM
Richard C. Stubben AMM
Osmond Smalley AOM
Paul S. Sperk AMM
Paul T. Sullivan AMM
Joseph Symbala AOM

Ernest H. Tessman AM
Emil Thau AMM
Grant G. Thompson ART
Samuel V. Tolston AOM
Clarence F. Vaughn AMM
Paul E. Vereruysse AMM
Rupert D. Westmoreland AOM
John J. Wienclaw AMM
Algur K. Williamson AMM
Charles E. Wolfe AMM
Perry J. Woodfill AMM
Mattelo V. Yacovino AOM

The Story of VP-63

IN MEMORIAM

These officers and men of VP-63 gave their lives in the service of our country:

Lieutenant James E. Breeding, U.S.N.R
Ensign Atvin L. Chambettin, U.S.N.R.
Lloyd Helming, AMM1c, U.S.N.
Howard S. Gordon, ARMIc, U.S.N.
George F. O'Cattaghan, ARM3c, U.S.N.R.
Henry L Bentz, AMM3c, U.S.N.
Lieutenant (jg) Henry Kovacs, U.S.N.R.
Jack H. Johnson, AP1c, U.S.N.
Hinam G. O'Dett, ARM3c, U.S.N.R.
Lieutenant Frederick A. Brown, Jr., U.S.N.
Ensign Robert D. Watten, Jr., U.S.N.R.
Louis W. Gratton, AP1c, U.S.N.
Guy N. Newton, AMM1c, U.S.N.
LaVetne D. Bentley, AMM2c, U.S.N.
Maurice B. Turnover, ARM2c, U.S.N.
Walter K. Abranz, ARM3c, U.S.N.R.
Eugene F. Harbour, S2c, U.S.N.R.
Lieutenant (jg) Billy E. Robertson, U.S.N.R.
William H. Golder, ACMM(AA), U.S.N.
Raymond C. Scott, ACRM(AA), U.S.N.R.
Arthur A. Rattel, AP1c, U.S.N.
Werdin O. Rude, ARM2c, U.S.N.
David R. Carmack, AMM2c, U.S.N.
Robert B. Law, AMM3c, U.S.N.R.
Lieutenant Woodrow E. Sholes, U.S.N.R.
Lieutenant (jg) Marvin S. Clinton, U.S.N.R.
John J. Kettmann, AP1c, U.S.N.
Walter E. Brown, ARM1c, U.S.N.R.
James O. Manning, Jr., AMM2c, U.S.N.R.
Wilhelm H. Watermann, ARM3c, U.S.N.R.
Edwin P. Christenson, AMM3c, U.S.N.
Lieutenant (jg) Robert I. Bedell, USNR
Ensign E.F. Horne
Ensign B.C. Wyrick
Charles Haupt, ARM

MAD Cats

Some of the stories may be repetitive; but each is seen through different sailors' eyes. Each story is its own, unique experience, something that goes with everyone else's experience to make a unique whole. For this reason, the stories are presented, as much as possible, in the words of the originator.

Sadly, some stories are missing; either because the individual was unable to respond for the book or was no longer with us. We acknowledge the part these MAD Cats played in the adventure. Many who did reply are sadly now gone also, taken by the unrelenting march of time. We salute them.

VP-63 Maintenance Shack Source: U.S. Navy

The Story of VP-63

PREFACE

Navy Patrol Squadron VP-63 was, indeed, a unique squadron having accounted for a number of "firsts" in Naval Aviation History Skipper Wagner and Hutchings relate the reasons for this in their *Story of Patrol Squadron Sixty-Three.*

A few words are appropriate concerning the aircraft that we flew. The *Catalina* flying boat was a very sturdy and reliable aircraft powered by twin 2,000 horsepower Pratt and Whitney engines. These engines were called upon to perform mammoth service in powering excess weights off the sea in all water conditions and keeping those massive hulks flying over thousands of miles with rarely a failure. As Commander Hutchings relates, having to lift and keep aloft 60,000 pounds was a severe test for an aircraft designed to carry a maximum load of 35,000 pounds. But they always came through in rain or shine and when the going got the toughest. Her flyers will always be grateful to the aircraft genius that designed and put this "darling" or "ugly duckling", as she was often referred to, together.

The normal complement of VP-63 consisted of eighteen aircraft and about 200 men. Personnel consisted usually of about 45 commissioned officer, 20 chiefs, including air pilots, 134 enlisted men including air pilots and civilian technicians. The normal requirement of a flying squadron of this size in equipment and technical know-how were tremendous. Added to this for VP-63 was the need to design, test, and maintaining new, for that time, and specialized gear which increased the normal logistics several fold.

City boys and country boys were brought together in one common cause to form the various units that fought the battles of World War II. Not unlike other wars or military units, VP-63 formed with men of all backgrounds including many that had never flown before the War. These men were asked to carry out dangerous missions often with insufficient experience for the task. Yet they did it with a high degree of excellence most of the time.

It was fun putting these stories together, although getting some of VP-63 personnel to correspond was like pulling hen's teeth. One person said it was one of the few letters he had written since the War.

MAD Cats

The 'spirit of 63' moved him off center. It was a real treat to talk to so many long lost buddies, but sad to hear of those gone on before us. I sincerely hope that this effort will bring back pleasant memories to those of us that remain.

Throughout this collection I have attempted to designate the rank and rates of the various authors coincident with the event. If I have mismanaged this affair by underrating individuals from time to time, I'm sure I will hear about it. On the contrary, if the opposite is true, I'll probably never be made aware of it.

I acknowledge with thanks all the VP-63 guys that answered the call to assist in this effort. Its fruition will be the result of their willingness to take the time, in an all too busy world, to look back to yesteryear. I, also, wish to thank in particular Skippers E.O. Wagner and C.H. Hutchings for their great history of the Squadron. Also M.B. Cummins for a copy of *Aircraft vs. Submarines* by Alfred Price from his library. Completion of this task proves once more our Squadron motto "The difficult we do immediately," and in my case, "the impossible takes a little longer".

I am grateful, also, for the great 1st and 2nd Squadron Anniversary books from which I was able to excerpt pictures and a few passages; also to Mr. Alfred Price for information gleaned from his article *Aircraft vs Submarine* and to Lieutenant E.H. Claypoole's *Cowboys Flew from Blitzville*.

Last but not least, we are all grateful to Andy Reid and Paul Ware for the idea of the reunion and getting it instigated by discussions with E.O. Wagner and to the "Kick Off Committee" for all their hard work in seeing to it that this 'forty-first reunion' became a pleasant and successful reality.

I'll always remember the *Ode To A PBY* submitted by John D. Linhart, AMM, to the Squadron's 2nd Anniversary Book *Along the Way*. It begins:

> Blessings on thee, you PBY,
>
> Staggering through the stormy sky;
>
> Struts and fittings loose and torn,

The Story of VP-63

and ends:
>And thy fabric loose and torn
>
>Every time I fly with you,
>
>In the misty night or morning dew;
>
>I know you must return with me,
>
>Or be with me eternally.

William P. Craddock
Bellaire, Texas
July, 1983.

* * *

This collection of stories from the men of VP-63, who lived and slept with the "Mad Cat" through almost three years of World War II, is not gory, like some war collections, because VP-63 personnel normally did not see that kind of war, the agony of injury and death on the ground and in trenches. Theirs was a clean war, usually over the beautiful sea and when tragedy struck, the consequences were usually instant and final. So if the reader is expecting this type of presentation, the wrong material is in hand. These stories are mostly humorous ones with tragedy generally in the background. In the main, the real life things that happened 'along the way' is what it is all about. (Editor)

MAD Cats

Lt. R.A. Barton

For some, as it did for many, it started at Pearl Harbor.

PEARL HARBOR: I was also on Ford Island in Pearl Harbor where Jim Hardy was on December 7, 1941. I was attached to VP-23. All our planes, except two, were damaged or destroyed by the first bomb dropped. During a relative calm between attacks, we managed to launch them. Our skipper, Massie Hughes, took off in the first one, and I in the second.

Being in the air seemed better than standing on a concrete ramp getting bombed and strafed. We were more angry than scared and were out to "get those bastards". We couldn't reach anyone by radio, and with no bombs and only one gun that would shoot - the starboard .50 caliber that would only shoot one shot at a time - our belligerence subsided somewhat.

We flew northwest on a search sector and about one hour out, saw an empty Richfield Oil Tanker steaming toward Hawaii being chased by a surfaced submarine. We dove to the attack (with no weapons) and, mercifully, the sub submerged. Everyone went on their way.

After patrolling uneventfully to our limit of about 600 miles, we returned to Pearl Harbor after nightfall. The only lights to be seen were from a burning ship which turned out to be the Arizona. As we turned on our lights looking for some water to land on, we were met by machine guns firing from many directions. Turning off our light we quickly flew out to sea to wait for our nervous gunners to calm down. Finally we came in low and dark through the entrance to Pearl Harbor and, seeing water reflecting light from the burning Arizona, landed and eventually found our ramp. Shortly after, several planes from Fighting Six were shot down while the poor guys were on final approach to the field. The following months were spent in the South Pacific participating in the Marshall and Gilbert raid (our first offensive action in the war), and a tour of the Aleutians. The Alaskan duty was interesting in that we used PBYs to dive bomb the Japanese landing on Kiska.

The Story of VP-63

Finally, I was given orders to return to the States to join a new squadron being formed in Alameda, California. I thought I had gone to heaven because this VP-63 outfit was supposed to spend the rest of the war looking for submarines off Trinidad.

I especially remember Bill Tanner's stories of starting the war by depth charging a sub to the north of Oahu just before the attack on December 7, 1941. With a certain amount of squeamishness, I remember Nick Nicholson explaining his bent finger with appropriate gestures. He was standing in the PBY cockpit yelling as he pointed to a missed mooring buoy - through the port propeller!

* * * * *

NEVER FLOWN A PLANE: Checking into VP-63, I met Captain Wagner who suggested that since we were already assigned a couple of PBY-5As, why didn't I round up several of our new Ensigns and shoot some night landings that evening. While walking out to the plane that night with my three "volunteer ensigns and crew", a fourth officer, a two-striper, came running up and asked to go along. The ensigns took turns in the seat, mostly shooting landings. It would be a pleasure working with these guys, they were really good and I marveled at their ability so soon out of flight school. As the last ensign left, the two-striper slid into the seat and I lined up with the runway. "I'll handle the throttles and radio. Tell me when you're ready. It's pitch dark over the Bay and there are no lights or horizon, so keep one eye on the instruments. We're doing 90 and about out of runway. Pull back anytime - WOW! You're half way to San Francisco. Turn in the pattern and we'll shoot another landing. Try a little more bank, you're skidding. Better line up with the runway before your final. Next time, flare a little before you hit the runway. Let's go around again." The second landing was slightly better, but I decided to call it a night. The PBY was a very rugged airplane. As we walked back to the hangar, I pulled the two-striper aside to suggest scheduling some familiarization flights. "Oh, I've never flown an airplane before. I'm an AV(S) assigned to VP-63. My name is Lyman Beardsley, Personnel Officer."

* * * * *

MAD Cats

SHORTER AND SAFER: Later, another instance, when walking by the Skipper's office after returning from a long, boring patrol out of Alameda, the Skipper was on the phone and he motioned me to enter. It was apparent he was talking to the NAS Commandant and saying that it couldn't be one of his PBY's that flew under the Bay Bridge while the Commandant was driving across because all his planes were secure on the ramp. When he hung up, it was my turn to explain how it was so much shorter and safer to fly under the Golden Gate Bridge and the Bay Bridge rather than brave all that soupy weather on top. Well, Captain Wagner could be absolutely eloquent when explaining how he preferred to have things done.

* * * * *

TEST PILOT: I returned to the States in January 1945 and was transferred to NAS San Pedro, California, as a test pilot. After being in PBY's so long, that too, was an adventure. Anything that climbed faster than 300 feet per minute and flew faster than 120 knots threatened to make me dizzy.

* * * * *

TECHNICAL ADVISOR TO MGM MOVIE STUDIO? Shortly after, I received orders to advanced base duty from BUPERS in Washington. I was to report to MGM Movie Studios to act as technical advisor in a Van Johnson-June Allison movie. Van Johnson was to ditch a PBY in the Pacific, rig a parachute as a sail, and end up on a beautiful South Pacific island where June Allison was waiting - just like real life! Our crew was the only one I know to try that in reality, but it didn't work. The plane sank. (Ed's Note: The truth of the matter is that Barton made a beautiful landing because there is a picture that shows the PBY riding high on the water with no damage after he was forced to land in the open sea because of a fuel line stoppage. The plane sank because of a storm.)

The Story of VP-63

Pembroke Dock May 1943. Source: RAF Coastal Command

*****

Lt. J.W. Hardy

<u>DECEMBER 7, 1941</u>: My personal experiences on this date involve mainly being on the receiving end of a lot of shooting and destruction. My squadron was based on Ford Island which sits in the middle of Pearl Harbor. I had the Duty on that day and was watching what I assumed to be the Enterprise planes making their usual Sunday morning practice air raid on Pearl Harbor. Suddenly, a little black dot dropped from the lead dive bomber and a P-boat hangar collapsed in a cloud of smoke and debris. It did not really dawn on us what happened until we saw the big red meatball on the side of the plane as it pulled out of its dive. Then torpedo planes heading for battleship row went by so close I thought maybe I could hit one with a rock. Our

MAD Cats

planes were un-flyable due to heavy strafing, which was lucky for us because we would have been sitting ducks in the air. We tried to shoot back with a nearby saluting battery, which was the height of futility as they only shoot blanks. I have had a very low opinion of saluting batteries ever since. We finally got some machine guns mounted on our busted-up airplanes and fired a lot of ammunition with inconclusive results. We had reports of Japanese troops landing at Barber's Point, so there were a lot of people around with guns and itchy fingers. That might, the stars looked like running lights on aircraft as clouds passed by, and whenever some scared kid fired a burst at them, every one on the island would open up. I think I was the noisiest night I never slept through. Some planes from the Enterprise made the fatal mistake of trying to land at Ford Island in the dark and our deadly (by that time) gunners downed nearly every one of them. That's about all I remember except that there was no liberty for about two months.

* * * * *

PRAYERS: Sometimes we had to take off at night from blacked-out seadromes. In particular, I remember Banana River where there was a single light in the middle of the takeoff area, but no other reference, and planes with no lights moored near the takeoff area. I prayed a lot on those takeoffs and afterwards I had a hard time believing my flight instruments that told me we really were right-side up. In Key West, we had to take off on a compass bearing between two blacked-out jetties. I prayed a lot then, too. Any pilot who has not tried to join-up on a lighthouse during a night formation rendezvous hasn't lived. I'll never forget the lighthouse at Pembroke Dock the night we left for Port Lyautey.

* * * * *

BLOODY GOOD SHOW: The seaplane ramp at Pembroke Dock was very steep because of the 20 or 30 foot tides and, any pilot overshooting his approach might very well lose the port wing tip float. It has been a long time ago, but I very clearly remember hearing about one of our planes in such a situation. Just before the port wing tip float was about to get wiped off, suddenly both wing floats retracted and the man on the wing shifted his weight out-board and the port wing slid

The Story of VP-63

smoothly across the ramp. Both floats then dropped as the ramp was cleared and the plane came around again for another, better approach. The Aussies watching thought it was a "bloody good show". I sure wish I could remember who did it.

* * * * *

RUN FOR YOUR LIFE: The thing to do at Port Lyautey was to go boar hunting; so I gave it a try one miserable, rainy morning. Someone stuck a shotgun in my hands and told me to stand in the middle of this trail through a cork forest, which was nothing really but a bunch of skinny bushes. I was to stand there while the Arab beaters chased the boars toward us. I was standing there, shivering in the rain wishing I were elsewhere, when suddenly I was face to face with a monster. I looked as big as a horse with beady little bloodshot eyes and big yellow tusks. I didn't think a cannon would stop it, much less a little ol' shotgun, and I heard how mean they get when only wounded. I looked around for a tree; but there were none big enough, in my opinion. I immediately reversed course at full speed and the boar did the same. The embarrassing part was that I was the only one to see a boar that day.

* * * * *

EXOTIC EGGS: One time I went down to the mess at Pembroke and saw a lot of people eating eggs for breakfast, the first time I had seen eggs since I left Iceland. With great anticipation, I walked up to the counter and asked a WAF for a "couple of eggs over easy". She just looked at me and I repeated the order. Finally she said, "But where are your eggs?" I found out that eggs were on the menu only if you brought your own. (Ed's Note: Everybody eating eggs at Pembroke, Jim? You must have dreamed this!)

* * * * *

MAD Cats

ANYBODY FOR A BUS? One time just about sunset, I was en-route from China Dry Lake back to Alameda with a load of Cal Tech civilians after test firing our retro-bombs. Unfortunately, there was no weather information available at Dry Lake, and we ran into lousy weather somewhere around Bakersfield. I couldn't even read the radio ranges through the static very well. We couldn't go back to Dry Lake at night (no lights), so I floundered around between Bakersfield and Fresno, low and slow, as my passengers became more and more apprehensive. After an hour or so, I found the highway and followed it into Bakersfield. I landed to check the weather and continue on to Alameda, if possible, but when I looked around for my passengers, they were gone! The had left without a word for Alameda by BUS! I learned that all the airliners between San Francisco and Los Angeles were held on the ground by Air Traffic Control that night until they knew I had landed at Bakersfield.

* * * * *

PRECIOUS RECORDS: Besides girls, there are two things Rollie Fairlie loved: bloody steaks and his phonograph records. Between Argentia and Iceland, we ran into some turbulence and Rollie immediately jumped out of the co-pilot seat and ran aft to sit on his cruise box full of those precious records. (Ed's Note: He also liked to drink good Scotch Whiskey, Jim, but what he liked even better was buying them for the British. It was rumored that he brought over a hundred of them one night.)

* * * * *

DECEMBER 7, 1941 REVISITED: Something happened in Coronado, California, several years ago that took me back to December 7, 1941 – and it was scary. During the filming of the movie **Tora Tora Tora**, a bunch of SNJ's fixed up to look like Japanese planes were flown into North Island to be loaded onto a carrier and hauled out to Hawaii for shooting the Pearl Harbor attack scenes. I heard the unmistakable drone of a large formation of planes overhead

The Story of VP-63

and I rushed outside to see what was going on. When I saw all those planes passing overhead with the big red meatballs on the sides, all I could thinks of was, "Oh no! Not again!"

I retired from the Navy in 1961, settled in Coronado, worked for a bank until 1970, and now I just play golf and tennis. Four children and the same wife, Jean, whom I met in San Francisco while we were at Alameda. I made a couple of trips back to Port Lyautey, one when I had a P2V squadron at Jacksonville and we deployed to Port Lyautay. It had not changed much.

Lt. A.H. Reid

OF RADIOS AND ANTENNAS: PROBLEM: Radio traffic VP-63 aircraft to NAS Alameda, "Base radio not getting through."

Scene 1: Intrepid VP-63 Communicator charged NAS Alameda Communications Officer with failure to man the operational circuits. Communications Officer replies, "The trouble is with your radioman. He doesn't know how to tune a transmitter." Intrepid VP-63 Communications Officer stomps from the scene, steaming from eyes, ears, and nose.

Scene 2: Intrepid VP-63 Communications Officer drops sheaf of paper on desk of NAS Alameda Communications Officer and speaks, "There's traffic from VP-63 aircraft last night. I copied it in my BOQ room, using a thirty dollar receiver hooked up to the window screen of my room." NAS Alameda Communications Officer to his CPO Watch Officer, "Better check our antennas and the antenna board."

Happy Ending: From that moment on, VP-63 aircraft that called NAS Alameda Base Radio received an instant response - day or night!

* * * * *

BETTER STICK TO FLYING, ANDY: Andy Reid and crew steaming blissfully into Jones Corner on a trans-Pacific, San Diego to Kaneohe, flight. Navigator to Pilot: "My Polaris shot shows we're 60

MAD Cats

miles south of track." Pilot to Navigator: "Nonsense! Take another shot." Navigator to Pilot: "My second shot shows we're 90 miles south of track." Pilot to Navigator: "I'm coming back and take a shot." Pilot in after-station lifts octant to his eye and both engines quit cold! Pilot and several others enjoy First Class Lurkey. Happy ending: Lots of wobble pump activity restores the engines. As they pass Mauna Loa off their port wing, receive a message from NAS Kaneohe: "Are you airborne?" Arrive at Kaneohe in good shape.

* * * * *

THE BATTLE OF THE LEMINGTON: The evening prior to departure from NAS Alameda, several bachelor colleagues awarded the Lemington Cross to Andy Reid. The obscene nature of the actual award precludes detailed description here; ditto, the achievements required to earn the award. For those not in the know, the Lemington Hotel in Oakland operated an Officer's Club for lonesome and wayward Navy airplane drivers.

* * * * *

THE FLYING BOXCAR: On a flight from San Diego to Alameda, the crew received a message from Alameda: "Alameda socked in. Return to San Diego." Query to San Diego received reply: "San Diego socked in. Try Bakersfield." Bakersfield advertised: "1200 foot ceiling; visibility 3 miles." So we dragged it through the weeds on a standard Radio Range approach and found the actual ceiling about 300 feet, visibility down to about one and a half miles.

Parked on the ramp, and as we were leaving the magnificent PBY-5A (a beauty she wasn't), an Army Air Corps Sergeant walked up to our plane captain and inquired, "Be honest now, did you FLY that thing in here?" Our plane captain walked slowly towards the Sergeant, in a kind of *High Noon* strut, and before anyone could interfere, the flail was on. There were no serious injuries - to the Sergeant, that is - and no charges were filed. (Names withheld by request.)

The Story of VP-63

SHOULD HAVE PUT IT IN OVERDRIVE, ANDY!: On the flight from the Salton Sea into Corpus Christi, we were cruising nicely at about 8,000 feet crossing the desert. The pilot was enjoying the scenery when an object on the ground sneaked into his side vision. It was the Santa Fe Super Chief and it was slowly passing the yoke boat! The Super Chief crossed the desert and passed out of sight, while the PBY ground her way slowly onward...ever onward. The PBY may have been a "low and slow" aircraft, but she was a bear for long-winded!

* * * * *

SHOULDN'T EXPECT TOO MUCH OF THE ARMY, ANDY: We arrived in Corpus Christi in less than ideal conditions for a ramp approach. Lots of breeze, heavy chop in the bay, spits of rain once in awhile, and cold. All of us made the ramp without incident and were tied up ready to get to the bar, when an apparition roared by, darned near on the "step". The apparition was an Army Air Corps Air-Sea Rescue PBY, enroute to base...somewhere. The pilot made several frantic passes at the ramp, always at darned near full throttle, as we watched, fascinated. On about pass #8, an object jumped - or was thrown from the after-station blister. Turned out to be the co-pilot in a Mae West. He paddled to the ramp and walked to us, dripping salt water. "My pilot says he needs help getting that thing on the beach. Anything anybody can do for us?"

Turned out the Army Air Corps aircraft did not know Navy frequencies, and hence, could not radio for information. Eventually the Army PBY was tied to a buoy in the bay and the crew ferried into the Air Station. That evening in the bar, I overhead the Army Air Corps pilot talking to a chance acquaintance: "I th'owed out both bags, but that didn't help, neither!"

* * * * *

SILENCE IS GOLDEN: Andy Reid to crew on what he thought was intercom after North Island requested full identification: "Millions of dollars in land-lines and radio equipment, and they want to know who we are!" Laconic voice from North Island tower: "You're on the air!" Talk about egg all over his face.

* * * * *

A PBY WILL FLY ITSELF, ANDY: The PBY in the Pacific Fleet in 1940 was equipped with SBAE (Stabilized Bombing Approach Equipment) that was also used as Auto Pilot. The SBAE gear was connected to the flight controls by an electronic clutch. Get the aircraft too much out of trim while on SBAE and the clutch would simply let go, resulting in either a severe pitch-up or pitch-down. Disconcerting to say the least, particularly in a sharp pitch-up. On our flight from Quonset Point to Bermuda, Andy Reid, the PPC, was at the navigation table when he noticed 'Shot" Lingle at his elbow. When he asked Shot who was at the controls, Shot replied, "We're on automatic pilot." Permanently branded by the characteristics of the SBAE gear, Andy Reid chewed on Shot sans mercy including, "Get your rear end back into the cockpit and NEVER do that again!" Forty years later, all Luis Corea could remember about Andy was how he chewed on 'Shot' Lingle that day.

* * * * *

ANCHORS AWEIGH? NEGATIVE: On the flight from Quonset Point to Bermuda, Andy Reid to Shot Lingle on the intercom; "Wonder what made that take-off so darned long?" Unidentified voice replies: "Maybe it was because you had both sea anchors out!" The sea anchors were long gone by the time we reached Bermuda.

* * * * *

THE ICE MAN AND HIS ICE MACHINE...ARGENTIA TO REYKJAVIK: On our night flight from Argentia to Reykjavik, the

The Story of VP-63

pilot was snuggled in fur flight gear, the heater going full blast in the Nav Compartment. (Remember the hot exhaust pipe of that heater? The one that burned large holes in trousers?) The cockpit was all cozy, warm, and sleepy. The PCC remembered the training film that admonished, "Frequently use the Aldis Lamp to check for icing when flying at night in icing conditions." He called for the Aldis, turned the beam on the port wing, and promptly had a No. 1, Class AAA lurky! (That isn't the only thing the Aldis was used for that night, Andy! - Editor)

The cabane struts were the size of telephone poles with rime ice. Use of the deicing boots stripped large sheets of ice from the wings. The thump when they hit the empennage was unsettling, to say the least. When the prop de-icers were on, the banging of the ice flung from the props explained for the first time the double hull in the prop track. All of this happened to a PPC who had NEVER encountered icing conditions in all his experience.

During our formation flight from Argentia to Reykjavik, our leader received reports from several aircraft that they were in icing conditions. He assigned several aircraft to fly at different altitudes to determine a better cruising altitude for the formation. Virgil Mayabb reported, "I'm at 8,000 feet and getting light rime ice." Then came the click of the mike leaving the air. It seemed like one or two seconds later that Virgil was back on the air, "I'm at 800 feet and it's raining down here." You'll have to ask Virgil how he got from 8,000 feet to 800 feet between clicks of the mike. He never told me the details of that one.

* * * * *

WAGGING HER TAIL BEHIND YOU: On arrival at Pembroke Dock from Reykjavik, we found the inner harbor confined for landing a PBY; confined and ringed with cliffs. On the final approach, across grazing sheep in a pasture, the PPC advised the crew on the intercom: "Make sure you're sitting , and get a safety belt. This one looks hairy!" Plane Captain Bradshaw, in a voice void of emotion, stated over the intercom: "The tail'll be right behind you."

MAD Cats

IT'S A MATTER OF A FEW FEET, ANDY: Enraged Welshman to Andy Reid after stopping Andy's jeep: "Get on the proper side of the road before you kill somebody, you bloody Yank!" Andy had been charging back to the squadron from E.O. Wagner's hospital with E.O. Wagner's signature on his detaching orders, orders back to Pensacola. On his exuberance, he had been roaring along on the right side of the Welsh road.

(Ed's Note: You mean the wrong side, don't you, Andy?)

Andy Reid's proposal for Official VP-63 Fight Song

(To the tune of *Dark Town Strutter's Ball*")

> OH HO, WE'RE HOT ROCKS AS YOU CAN SEE
> CAUSE WE'RE THE BOYS FROM SIXTY-THREE
> WE'VE GOT PLENTY ON THE BALL
> IF YOU'LL GIVE YOUR NUMBER WE WILL CALL
> AND IF YOU LIKE, WE'LL GET YOU TIGHT
> AND IF YOU CARE, THERE'S A CHANCE WE MIGHT
> SO PLEASE DON'T STRUGGLE, AND PLEASE DON'T FIGHT
> THERE ARE PLENTY OF OTHER GIRLS IN SIGHT
> AND WE'VE GOT LIBERTY ALL TOMORROW NIGHT!

> (Repeat ad nauseam)

Anonymous (at this time).

* * * * *

Lt. R.M. Bush

BACK FROM THE SEA: We took off from Alameda and came out on top at about 5,000 feet. We went a few hundred miles when we saw a large front up ahead. Lt. Breeding decided to fly over the front since we had seen a P-38 coming over it going to the south. We climbed to 10-11,000 feet and had to put on re-breathers. We had not gone too long when we picked up clear ice. We had picked up so much clear ice that shortly thereafter, we went into a spin.

I don't know what the other crew members thought at this juncture, but I thought we had had it. We kept spinning and finally came out at

The Story of VP-63

5,000 feet, between two layers of clouds. We had lost all antennae and could not raise any bases. We finally saw a hole in the clouds and circled down and saw a small town next to a river. We identified the town through a sign on top of a diner. We could not climb higher than 5,000 feet because of carburetor icing. Lt. Breeding had ordered "pre-heat", but someone didn't get the word. We tried to get rid of this by backfiring, but with no success. We then decided to take a course between two mountain ranges which were at about 5,000 feet. We finally had to let down and sighted the ocean and located ourselves of Hecata Head, Oregon. We dropped out a trailing antenna and picked up the Seattle base. We were out of fuel and they told us to land at sea and they would send a boat from the Coast Guard station at Empire, Oregon.

The MAD Boom. Source: U.S. Navy

The weather was foul and we could not get enough altitude to drop a float light and see where we were going to land. Therefore, we took a blind chance and, unfortunately, landed in a trough and thereby tore a large hole in the bow of the aircraft. Lt. Breeding suffered a bad wound to his forehead and was slightly dazed. Water was coming in very fast so we got back into the engine compartment and closed the

MAD Cats

watertight door. We put a two-man raft out of one blister and a seven-man raft out of the other. Shortly thereafter, the plane sank and Burtz and I got the two-man raft.

This was the night of New Year's Eve 1942 and this experience was the most traumatic one I have had to date, forty years later. We crashed about 5:00pm and, after much confusion, Burtz and I drifted away into the rough sea. I remember that he was dropping the flashlight into the bottom of the raft, but I had no perception that he was freezing. I told him that I heard surf pounding and to be sure to grab the lifeline when we hit the breakers. Not too many minutes later, we both went sailing into darkness.

Luckily for me, I had bunked in Coronado with a fitness nut named Cheverton who, each morning, made me join in riding the surf. This knowledge came in good stead since I was able to ride the surf all the way into the beach. The beach at Hecata Head, Oregon is most inhospitable, since it is nothing but rocks covered with barnacles. The waves kept throwing me against the rocks and then dragging me back out again.

I sighted my watch and it was exactly midnight. I was exhausted and felt like giving up, my mind switched to the previous four or five New Years Eves which I had spent in a small Connecticut town. The custom in that town was for people of all ages to go to the school auditorium and socialize until midnight. At that juncture, one of the local ministers would say a prayer and everyone would join in singing a hymn. After that, everyone celebrated until early in the morning.

All of the above probably took only two or three seconds to recall. It was then that my feet hit bottom and brought me back to reality. I got a renewed desire to get ashore. I was finally able to hold onto one rock and gradually crawl up to a spit of land. I was found by Chief Don Tuttle and Seaman Dave Rodgers of the Coast Guard.

Some years ago, my wife and I were invited to a formal New Year's Eve party. While looking for my cuff links, I came across my Navy wings and decided to wear them to the party. A man I knew casually, Tom Moore, came up to me and started questioning me as to where I had been in the Navy. It turned out that he had been duty officer at

The Story of VP-63

Whidbey Island that New Year's Eve in 1942 and was sent down the next day to conduct the survey. Small world – what??

Cmdr. C.H. Hutchings

<u>ALL ABOARD</u>: In October 1943, Adm. Sample and Com. Hamilton visited VP-63 at Pembroke to discuss strategy, tactics, logistics, and operations of the squadron. When it came time to depart, the MAD Cat CO escorted them to the train, and boarded it with them in order to continue the discussion as long as possible. He thought he would get off as soon as he heard "all aboard". There was no "all aboard" given, and suddenly the train started to move. He rushed to the door, but it was locked, Com. Hamilton said, "Hutch, you might as well relax and enjoy the ride." Two hours later, the MAD Cat CO was able to get off and phone the squadron Executive Officer who was about to instigate a search for the missing Skipper.

* * * * *

<u>NOW WE KNOW!</u> At Pembroke Dock, the tea bell frequently rang when our aircraft were being launched, interrupting the launch. Finally I complained to Group Captain Carey, the base CO. He called in the Base Medical Officer and asked him if going without tea would be injurious to health. Upon receiving a negative reply, the Group Captain issued an order stopping morning and afternoon tea on the base. I kept very quiet about the incident because I didn't want everyone mad at me. The interesting part of the story is this: without realizing it, I had become hooked on the tea habit and cancellation of tea hurt me as much as anyone else. (Ed's Note: Just being mad, Skipper, is a mild way to put the feelings that would have ensued if we had known you were the culprit. Besides, it wasn't so much the tea, it was the crumpets that kept us one step ahead of anemia.)

* * * * *

MAD Cats

Lt. M.W. Nicholson

ON A DARK AND STORMY KNIGHT – A LARGE FLOATING ROCK: Pembroke Dock, 1943. It seems there was this fearless and most experienced crew called to launch a provoked attack on the Hun subs in the Bay of Biscay after shooting down as many JU-88s as they had time for while proceeding to the target area. It was about 3:00am when this sterling group of pilots, mechs, and radiomen manned their STEED tethered at a buoy in the upper reaches of the Milford Haven estuary. The first event to be performed was to taxi their Monster to the lower reach of the river to get a long takeoff run. As it turned out, this was no small feat to joust with a crooked river with the tide at low ebb! To make a long story longer – even though the dauntless AP was guiding the idiot at the controls with an Aldis Lamp, it seems that for no good reason A LARGE FLOATING ROCK decided to find its way to this valiant steed's bottom! It couldn't have been over one microsecond before nine of the ten persons embarked announced quietly to the pilot that their floating palace was sinking. The events to follow are replete with valorous and heroic acts of the US Navy crew fantastically assisted by the Royal Air Force boat basin personnel, fire-boats, and many Sunderland crews moored in the way of the foundering PBY trying to make its recovery ramp!

For those familiar with the normal damage control procedure of sticking pencils into popped rivet holes, you will have to forget this notion as the radiomen and mechanics found: 1) that there were not enough pencils, and 2) that telephone poles were sorely needed for this normally routine function. So as not to be thwarted in this survival situation, this imaginative crew put pillows, mattresses, and an occasional life raft in the slightly oversized holes.

Meanwhile, back at the pointed end of the boat, an RAF bumboat was being secured by a tow line to the snubbing post! At last we were on the way to salvage for this sterling steed. It was just getting light when this 'no-panic' crew noticed that the RAF boat had us underway for the ramp area – unfortunately as events to follow would show, the RAF coxswain thought that a 100 foot tow line was appropriate. With this length of line, the ruptured steed was swinging left and right of the rear end of the towboat in uncontrollable oscillations. Someone woke

The Story of VP-63

up the pilot to this impending disaster just as the left wing-tip float cleanly cut off a Sunderland's float! In transiting further up the RAF mooring area, it was noticed in passing that this incident must have been exactly timed with that Sunderland's reveille. As a number of people appeared topside, some must have had poor footing and went over the side. Our charger was still oscillating beautifully and we were fortunate enough to witness a similar reveille on another Sunderland that was on our opposite side. On close examination, it was missing a wing-tip float also!

Bear with us. This saga is approaching an almost uneventful end for the beaching ramp was now close aboard! It was noticed by Sandy, or someone else with 'eyes of an eagle' that the S.O.B. (Sweet Old Boat) seemed to be riding quite low in the water. As a matter of fact, there seemed to be a little water lapping over the Blisters, but not quite up to the CABANE STRUT window! Some farsighted devil thought it would be a good idea to bring the RAF fireboat alongside to pump out a slight amount of water (quite rapidly) so that this Steed wouldn't drown immediately and thereby block the launching ramp 'til the end of WWII.' (You see, if that had happened, it would have meant "liberty" for all hands for some time – and nobody wanted to go home without being shot down at least once.)

By now, things were going really beautifully. The fireboat was nestled alongside, pumping like mad, and the Charger was rising with its stirrups out of water! Suddenly, the normally tranquil Britisher in command of the fireboat was seen to be quite exercised while dashing madly to the blunt end of his boat where the pumping valves were! This action was not understood by the Steed's recalcitrant crew, nor for that matter by the handsome beachmaster who was smiling because the Beaching Gear had finally been attached – and the rescue operation had ended successfully!

It seems that there was more than 'one nut loose' that early British morn! The seaman in charge of valving in the fireboat had forgotten to place that little pumping valve in the "Overhead Discharge" position. So instead it (the Steed's water) poured into the fireboat bilge – and heroic vessel FOUNDERED! It was a "bit of a sticky wicket" for our Captain 'Crackers Carey' for some time hence! On

MAD Cats

the other hand, the *Sterling Steed's* crew was lauded for their actions and awarded double duty.   LOVELY!!

Harold Irby, ACRT

TOAST THE BRITISH SOIL: Shortly after the Squadron moved to Wales from Iceland, one stalwart had fetched along a good assortment of 'fifths' and in the barracks corner the reunion was toasted. "Hooray tonight. . . headaches tomorrow. . . should have drowned me instead of my sorrow." Result: the host had a half dozen fifths left over and a king-sized hangover. He uncorked each bottle, unbeknownst to any, and doused the British soil 'til nay a drop remained, vowing never to again imbibe. The next day he was out sniffing the remaining fumes from the soil, wondering how he could get the squeezings back in the bottle.

* * * * *

WATERMELON SHINER! Often in late afternoon, flights of B-17s came back from raids on Germany over our base, still in battalion formation...here and there one missing. As one flight approached, a cripple lost altitude and made an emergency landing in a farmland nearby. Non-flight VP-63 personnel skirted the hedge rows in ammunition carriers to the rescue. They arrived to find the B-17 crew circled around a gunner crew member with blackened eye. What gives? The flight was the test run from England via Regensburg, Germany to bomb the ball bearing industry, and then on to Africa for landing. They would reload and bomb the same target on their homeward flight. Much flak...much evasive action. The gunner was the only one injured on the B-17. Apparently, he had worshipfully concealed a watermelon to enjoy back in England, and it had become dislodged during evasive action and hit him in the eye giving him a Bourbon Street shiner. No sympathy...just a corny suggestion to "duck faster next time."

The Story of VP-63

<u>CHIC SALES</u>: I remember the day Nick's crew rotated home. Native outhouses were of flimsy construction and sitting ducks for the air stream from a PBY at 75 feet. "Not nice" read the Skipper's directive to all pilots to cease collapsing these structures. Bets were made Nick was going to have a parting shot at the village privies as he slithered up from the Wade Sabu River at Port Lyautey. He did a 180 and...you guessed it...took aim...gave a little altitude and throttled the bird. Nick, we never knew whether on your goodbye salute to us guys remaining, you toppled a couple. Tell us!

* * * * *

<u>BLACK PLAGUE</u>: Vaccinations...Cork off...Excuses...anything to avoid them. Once it was different. The Black Plague was in a village less than 50 miles away. There was no known cure...you agonize and die. It took just one "Now hear this" and a 100% turnout trotted to sick bay for that shot. Surely you remember that most painful, aching injection...but we lived to see another day, so the medication did its job.

* * * * *

<u>MAIL RUN</u>: I remember watching the DC-3 on mail run to Malta and other points loading up on cigarettes on a couple of occasions (to peddle on the black market) and leaving bags of mail at Port Lyautey for the next plane to take off.

Says Harold Irby, ACRT, whose business experience prior to WWII included work in a New York advertising office and, since then, in real estate and appraising at Winter Haven, Florida: "Why I'm going to San Diego for the reunion. That was the finest group of men I've ever been associated with. There was a rapport among us all other squadrons envied. We were a happy unit. We knew what our jobs required and every man had the self respect to do his job 110% and lend a hand when a mate was overloaded. God blessed so many of us in close places, allowing us to survive. Many a tear was shed for each one of our team who was lost."

* * * * *

MAD Cats

(Photo: VP-63)
BOQs at Port Lyautey

W.E. Foreland, ACRM (AA)

DOES HE ALWAYS FLY LIKE THAT? In a moment of light-headedness, I was talked into taking a cruise down to Junkers Junction by CAP Tony Mule on the premise that as Radio Chief, I should check out the MAD gear in their aircraft. Crew 8 was known to be a bit flaky, being captained by Lt. Nicholson (one of the original cowboys from PD) while Lt. Russell and Ens. Horne tried to compete with their noble leader, CAP Mule, who was continually wisecracking and flew just as funny as his Brooklyn accent. The enlisted members of the crew had gained their reputations ashore in every town near any air base where VP-63 stopped long enough to pull on their dress blues. This patrol was particularly peaceful and calm, although the rarely seen clear skies kept everyone alert for JU-88s. As the nearly empty PBY-5 approached the coast of South Wales, she seemed to float on the air with her engines purring.

About the time we came into view of Milford Haven, Nick saw a lumbering Sunderland flying boat, no doubt an Aussie crew from 228 Squadron, lining up for a long glide into Pembroke Dock landing area.

The Story of VP-63

Nick passed the word to buckle up, poured on the power, overtook the Aussies, and made a steep diving bank to starboard.

Belted to the navigator's seat, I could look out the small window on the port side. As the aircraft came to level flight, I saw some row houses approaching very rapidly, with a chimney passing between the float and me. Then, after we trimmed a hedgerow, I saw this Welsh pony galloping alongside us with the prop-wash blowing his maine.

Bracing myself for anything, my next sensation was a prop hanging full stall that I knew would pop some rivets when we finally hit. Instead, the old bird just flopped onto the water, shot across the estuary in front of the Sunderland, made a complete turn in front of the ramp, and was beached without a hitch.

As we climbed down on the ramp, I got my voice back and asked Tony, "My God! Does he always fly like this?" He looked at my white lips and calmly asked, "What way?"

* * * * *

NOW HEAR THIS, BILL HAMILTON! After a somewhat contrived delay of two days at Quonset Point, Crew #5 was finally ready to join the small contingent of VP-63 PBYs on Bermuda. All the gear for advance base operations was aboard, including beaching gear, making the old bird sit low in the water as she taxied out for the takeoff run down the bay.

Lt. Andy Reid, the PPC, swung the plane's nose into the wind and pushed the throttles forward as Ens. Van Lingle adjusted the props. Number Five usually jumped on the step quickly; but, this time she couldn't seem to get moving. With RPMs at the extreme limit, the speed began to build, and with a brisk head wind, we almost became airborne.

Keeping the nose down until we passed under the Narragansett Bridge, Andy pulled back on the yoke and we broke free of the water. Everyone knew we weren't that heavily loaded and when we looked back, we saw two sea anchors streaming in the wind. They had evidently been stowed on the fuselage behind the blisters and were blown off as we taxied.

MAD Cats

The air became blue and then purple as our trusty commander let the crew know what he thought our stupidity. Poor Bill Hamilton, who happened to be sitting in one of the blisters at takeoff, receiving the brunt of the dressing down. It shows to go that there ain't nuthin' you can't do with a PBY

* * * * *

BELLY CHECK: During the time VP-63 was working out of Quonset Point, some of the men came down with catarrhal fever (flu to you) and were sent to sick bay at the base hospital. While they were there, one of the other inmates broke out with a case of scarlet fever. The whole ward was immediately quarantined by moving the uninfected sailors to an unused wing.

The wing had a door at the end that opened onto the walkway leading down to the squadron work area. Good Samaritans, like Red Bradshaw and Ray Scott, decided to bring the mail around to those poor prisoners.

After a few days the routine was established and the schedule for 'belly checks' was set. This allowed the messengers to stay for a game or two of cribbage without being caught. Except for one time when Ray Scott overstayed the time and the retinue of corpsmen and doctors came through and examined everyone's belly, including Ray's. Of course, he left immediately after the examination.

About 20 minutes later, a corpsman came back and made everyone line up and give their names. When he protested that there had been an extra belly, he was told he was crazy and laughed out of the room.

* * * * *

SHOWER HOT, CHIEF? VP-63 was a squadron of real characters, but one of the most colorful was Bill Golder, ACMM. He took a maniacal glee in flushing all the toilets in the Chief's head when someone was taking a shower. When the victim screamed at being scalded, he would whoop (manaically)

The Story of VP-63

B.F. Pfannstiel, AMM

<u>RIP VAN WINKLE</u>: On this particular trip from Alameda to North Island, Laurence Hickey was the Plane Captain and I was the second Mech. Hickey came in that morning, as he used to like to call it, "is usual charming self", If anyone remembers, he could be like a zombie when he was really loaded.

Somehow he got the pre-flight done and I was glad the planes were new. He got into the rear port side bunk and was asleep in two seconds. He was still sawing logs when he got to the chocks at North Island. I jumped out of the tower to wake him before Lt. Kauffman, who was PPC, came back there. I shook him awake and he looked up at me with those bloodshot eyes and said, "When we are going to take off, or has the hop been concluded?"

* * * * *

<u>DETAINED BY THE CITY</u>: Then there was the night in 1943 on detachment at Jacksonville from Quonset Point that George Sips, McCrea, and I were guests of the City for a night. It seems that we were talking too loud in this hash joint before going back to the base. The next thing we knew, the SPs had us. They were rough in Jax at that time. Ex-cops!

The bunks were made of ½" boiler plate with holes for air drilled all over them. The holes were about 1" in diameter and the bunks had no mattresses. Sipe and I had on whites and we looked like zebras the next morning after lying on those bunks all night. McCrea had on khaki clothes and his apparel didn't look so bad, but his health looked awful. We were worried about him going back to the base in the back of the paddy wagon with the temperature approaching 120 degrees. The result of this little sortie was that we got restricted to the base for a couple of days. Lt. Hardy was in charge of the detachment and we got off light.

MAD Cats

ALKY: There was the time in Iceland when we were running low on 190 alcohol for the prop de-icers. No wonder, someone was drinking it!

* * * * *

UKELELE IKE: The there was the great *Ukulele* player, Bruce Mithee. He slept on the bottom bunk under me at Pembroke before he made the hat. He used to serenade us with that thing every day. I wonder if he ever learned to play it. But, I'll have to say a good thing about ole Bruce. He had a couple of fast fists. We sparred a few times and he taught me a few tricks. I guess Harry Beam is the only one I could whip. He was a sucker for a left hook. No defense at all! There was also the time, I think, Bruce had a misunderstanding with someone waiting for a dinghy at the boat dock and somebody wound up in the water.

* * * * *

WELL ARMED, EH? We were coming back from patrol in the Bay when we spotted a German seaplane, a Dornier 26. He was off the starboard bow and possibly a mile away. Moose Ferris was either in the bow turret or in the co-pilot's seat, and upon spotting him said, "Let's go down and knock him off." There was no engagement with him and probably lucky for us. We learned on return to the base that the Dornier was very well armed and not one to be fooled with, especially by a PBY armed as I was.

* * * * *

DECEMBER 7, 1941: On this date I was in VP-43 on North Island lying in my sack reading the Sunday morning paper. All of a sudden the MAA came running through the barracks beating everyone's bunk with his club. He was hollering that the (Japanese) had bombed Pearl Harbor. We didn't believe him at first. He finally got us up and we mustered at the hangar right away. Planes were armed and patrols started that afternoon. I was on the beach crew and we worked until Tuesday before I was able to get some rest.

Lt. M. Kauffman

OFFICIAL REPRIMAND: VP-63 brings to mind all the development and experimental work and flight operations on both coasts, Iceland and in England. I cannot recall why we were in Bermuda (convoy escort, I presume), but on May 9, 1943, I landed in the open sea (rough) and picked up four British survivors who had been drifting for 39 days (torpedoed off the west coast of Africa). After dumping all loose gear and half my fuel, and with the help of God, got airborne. Upon returning to base, I got an official reprimand for landing in the Atlantic, against all regulations (here, but) routine in the Pacific.

* * * * *

MASCOTS ALONG THE WAY: On July 20, 1943, the final detachment of VP-63 arrived at Pembroke Dock, Wales, from Iceland. The following morning a crisis arose. Carefully hidden aboard planes that had arrived in the second detachment were three VP-63 mascots, dogs of questionable background. Answering to the names of Snippsie, Schnoppsie, and Sniffer, they had been penned up since leaving Iceland 36 hours before. All were unhappy and said so in howling terms. Even numerous ferry trips with the dinghy loaded with food, water, and a crude miniature fireplug failed to placate them.

Their flying owners had heard of the laws of the UK. Dogs could not be imported under any conditions, whether it be for companionship, to build morale, or for any similar purposes – and that was that! Several attempts to smuggle them ashore failed. Each time the dinghy containing the pets neared the ramp, up would pop an arm of the law and back would go the dogs. The situation grew more and more impossible.

As yet, the actual presence of VP-63's pets was unknown to local authorities. Their intermittent cries of anguish had been attributed to animals on the opposite side of the Reach. However, Snippsie, Schnoppsie, and Sniffer were not aware of this, or by this time they

MAD Cats

just didn't care. When the third attempt to make the beach failed, it was decided to place them all in one plane, the one moored farthest away from the base. This was adding insult to injury, so the pups put their heads together and howled in unison. Hatches were closed as quickly as possible. Both blisters were secured, but still the noise persisted. Finally, quick thinking Lt. Sam Parker, USNR, the Squadron-style dog doctor from Montana, and pilot of no mean ability, hit upon an ideal solution. Since we couldn't stop their noise, why not drown it out? In less time than it takes to tell it, he started both engines of the PBY. They needed a test run anyway, and within 30 minutes, all three dogs gave up. Even today you can get an argument from the boys as to which makes more noise – two engines or three dogs.

Meanwhile on the beach, a council of war had been called. Share holders in the problem pets – and that meant half the Squadron – were in deep conference with a scholarly Yeoman who somewhere had dug up a law book. He was reading aloud page after page of assorted findings, opinions, interpretations, and reviews. Torts were greeted with retorts and then it happened! A case was discovered that dealt with disputed property. It listed the evidence presented to support the winner's contentions. This was it! Later that day, Snippsie, Schnoppsie, and finally Sniffer came ashore. As the landing dinghy hit the ramp on each trip, the identical challenge of a progressively more irate official was met with a bland smile and an indisputable document. Oh yes, sir! This dog was purchased just recently. Yes, here on Wales. See? It says so right here below the tax stamp on this genuine bill of sale. The now happy pets soon were sniffing and smelling their way around the landing ramp. Almost in rage, one Army major exploded, "Great balls of fire. What an outfit! Yesterday they took over the station! And now they're running the Civil Government! Will someone please tell me what in the hell you cowboys are going to do next?"

Doc Spears and Sniffer were as inseparable as a man and his trousers. Sniffer had over 500 hours of flight operational time at the end of the first year of VP-63 existence and had been aboard when Doc and crew sank a sub. Sniffer was recommended for an air medal, but had to settle for a dog tag. It was routine for one of the crew to take Sniffer

The Story of VP-63

up on the wing after each patrol and then meticulously wash down the upper surface of the wing. Lt. Gilbert Knight, who flew throughout the Bay of Biscay Campaign with Doc, says that Sniffer was always nervous about flying in the Bay. (Ed's Note: So were a lot of others, Gil!)

Squadron Mascots. Source: U.S. Navy

For Schnoppsie, the monetary values assessed by man meant nothing. He had come through purchase into the Squadron and the dachshund never became "browned off" so long as three meals a day were provided. Brooks had nursed him through months and thousands of miles of flying. But once in extra curricula diversions, the "cubes" frowned on Brooks and his till was soon empty. In desperation, he gambled, "Shoot Schnoppsie against two bucks"; as the dice stopped on an ace/deuce, the beagle exchanged ownership. He continued on with the Squadron and became one of the first mascots to fly the Atlantic.

The stay in "Blitzville" (Pembroke Dock) brought an increase in the number of pets carried on the VP-63 Roster of Mascots. Snippsie,

MAD Cats

Schnoppsie, and Sniffer were augmented by such characters as Mickey; Dashing Dilbert; and the more sedate Pembroke, a Sealyham; Radar, a gentlemanly English Setter; and Buddie, a much beloved Cocker Spaniel. When Schnoppsie, with better than 300 hours of flying time to his credit, developed bad ears and had to be grounded, it was the young aristocrat Pembroke who took his place in the flight crew. Actually, the dogs were common property of the Squadron, although in theory each had a separate master. Perhaps the common property designation came as the result of ever mounting lists of misdemeanors charged against them. Once, when high ranking British and American officers came to Pembroke Dock, the irresponsible Buddie persisted in leading a barking bank of capering canines right on the heels of the official inspecting party. To the delight of squadron personnel and the mortification of the American and British resident commanders, efforts to drive them away brought disdainful barks and well feigned growls of anger. They say every dog has its day, and the pets of VP-63 really made the most of theirs.

The following morning brought semi-official repercussions. It took considerable maneuvering; but as was inevitable, the whole matter adjusted itself and the dogs remained. However, all except Radar were subjected to a Summary Court Martial embodying all the dignity and seriousness of true Navy Justice. Buddie, as the ringleader, was sentenced to loss of flight status and remunerations and, along with all but one of the rest, was deprived of his weekly ration of frozen meat scraps. The exception was Snippsie. She got off without penalty, some say, because of having been merely an innocent bystander who went along to see what was happening, while others claim it was the testimony of the Flight Surgeon who stated that a lady in her condition, in a family way, could not be sentenced to lightened rations. Radar, the only mascot who clearly had not been party to the episode, was given a commendation for his dog tag and excused from the weekly bath for a period of one month. (Ed's Note: Sniffer passed away in 1950. Don't know about the rest.)

The Story of VP-63

Dashing Dilbert

Mickey

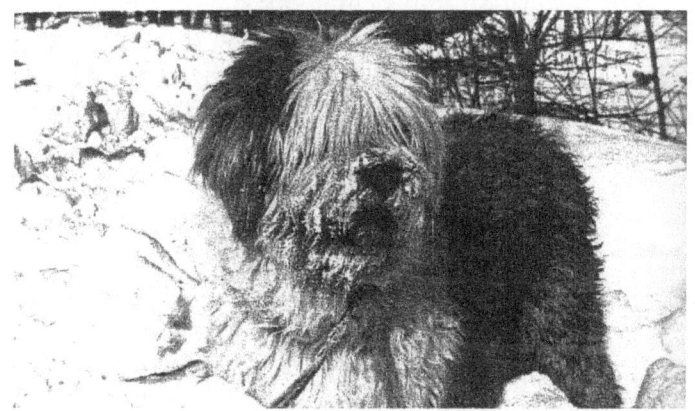

(Photos: VP-63)
Pembroke waits for the mailman.

MAD Cats

Lt. V. Mayabb

WHAT WAS THAT? FLAT SPIN A PBY? Spinning a PBY in the middle of the Atlantic, in the middle of the night, in the middle of a storm - who else but Virgil Mayabb and his crew? On our way from Argentina to Iceland, when we moved from Stateside to England in 1943, we were cruising at 9,000 feet in a spread formation of three PBYs. Sully Kauffman was leading the flight, Jerry Korelock was flying No. 2 position, and Mayabb the third. We had our auto-pilots engaged. Then all of a sudden, the yoke came back and hit me in the chest. By the time I disconnected it, we were in a flat spin. When I finally got control of the aircraft, we came out at about 500 feet. Well, I got her flying straight and level again and climbed to 1,000 feet. I then called Sully and told him the weather was much better down here. He told me to continue to Iceland on my own. We arrived safe and sound. Then after replacing a bunch of rivets, we were ready to proceed to our destination, still singing our song, "We're too young to die in a goddam PBY." (Ed's Note: And now you know, Andy Reid, how Mayabb went from 9,000 feet to 500 feet between clicks of the mike. It must have been that darned that darned SBAE equipment you were talking about. Reckon it got too much out of trim, Andy?)

* * * * *

MR. SILVER EAGLE: For the benefit of those who didn't know, Virgil was awarded the "Mr. Silver Eagle" Award at the 12th reunion and convention at Norfolk, Virginia, in October, 1976.

* * * * *

MORE ON THE FLAT SPIN: Here is an expansion of the flat spin from Howard Lee, CAP, who was in the co-pilot seat that night when the auto-pilot malfunctioned. "We were in the soup when the yoke suddenly came all the way back and could not be overpowered. The P-boat stood on its tail. I reached up and cut off the auto-pilot and Virgil nearly shoved the yoke through the instrument panel. We fell off and did about one turn of a spin before we finally got things back

The Story of VP-63

under control. The plane was a mess. We had ten men, all their baggage, a bicycle or two, and other miscellaneous items strewn all over the planes. The Plane Captain was heating soup at the time and it went all over the compartment. Everyone not tied down had a wide assortment of nicks and bruises, but we were still flying and nothing seemed to be busted, so we all had good reason to rejoice."

Flight crew gets last minute brief before a mission. Source: U.S. Navy

E.F. Bradshaw, ACMM

IN THE WRONG PLACE? V.L. Robinson, ACMM, and I were working one afternoon at Pembroke Dock and one of the warrant officers in the Maintenance Unit took us to the BOQ for a drink. It was Thanksgiving. Well, one drink went to two, etc. I was going down the passageway looking for the head and ran into Capt. Hutchings. He asked me how my nose was as I hade recently fallen on the ramp and broken it. My reply was, "In the wrong place tonight, Sir."

* * * * *

CHIEFS GO BY SHIP? When the Squadron moved from Wales to Port Lyautey, Robbie and I were ordered to ride the ship. When we were loading equipment, someone called from the bridge. It was our Senior Aviator from the *USS San Francisco*. He was Captain of the ship and welcomed us aboard - no duties! It sure was a nice trip, thanks to those orders.

E.H. Tanneberg, AMM

THAT GLORIOUS STINKING WEATHER IN THE BAY: Lt. James W. Hardy, USN, at that time Executive Officer of the Squadron, put it best: "The worse the weather, the better we like it." The native Oregonian's reasoning, of course, was that bad weather meant clouds and lots of them - and clouds meant protection from German fighter planes. Everyone remembers Lt. Benscoter's message to Base, "Weather stinks." A WAAF officer, who decoded the message before relaying it to our Operations Department, exercised her Elizabethan reserve, and couldn't resist massaging "Big Ben" : "Stinks less farther west!"

The weather was anything but glorious the night Lt. (jg) Parker tried to get back into the Base after a 12-hour flight in the Bay. An entry in the flight log of AMM 1st Class E. H. Tanneberg, Plane Captain of the crew headed by Lt. Parker, adequately explained the weather they

The Story of VP-63

encountered that night. Lt. Parker and crew had experienced very bad weather all day. Upon returning, they learned by radio that fog covered nearly every known landing area. They were directed to try a major English harbor previously designated as unsafe for any and all seaplane operations.

Lt. Parker flew through fog, thick fog, the kind you expect the propellers to emerge from with dull edges. For three hours, the PBY wandered over and around the harbor of that English seaport. Twice they pulled up sharply to avoid sprawling, blacked-out warehouses.

Once they saw a clear area - saw just enough of it to show they were flying below the loading arms of huge dock cranes. Somehow, though, they made it down and felt their way to a mooring. That night a God-fearing, flying sailor from the state of Washington, logged his flight time for that day with, "Flew twelve hours with Mr. Parker; three hours with the Lord."

Lt. (jg) M.J. Vopatek

<u>NOSTALGIC MEMORIES</u>: Memories are made of these, so saith Mike Vopatek.

Being stationed in Iceland in July, the sun shining for 24 hours a day. As Lt. Sam Parker aptly described it later on, "The only way to know it was time to go to sleep was when they closed the bars."

In England, nine of us being stationed off the base in former married officers' quarters and the "glad to see you back" ritual in which an ample supply of bourbon whiskey in a tumbler, mail from home, and a glowing fire in the fireplace was ready for those of us who returned from that day's patrol in the Bay. (Ed's Note: Where did you get the fuel, Mike?)

In Africa, having Joe Louis put on a boxing exhibition during a three-day visit. Then a few weeks later, having the opportunity to talk to Humphrey Bogart and him signing the "Short Snorter" bill.

MAD Cats

Lt. (jg) R.M. Fairlie

OF PIGEONS: Do you remember the pigeon loft opposite house #7, quarters for a few of us up the hill a bit from the base at Pembroke Dock? How we used to get "Pidge" (in charge of pigeons) a little "charged up" with alky and grapefruit juice – and when he was real gone, we would 'charge' the birds, too, and then set them loose for their exercise and watch the poor birds spin in, fly into each other, and into the loft attempting to return home? The things we did!

* * * * *

RABBITS: Do you remember the crew going out on patrol that day at Pembroke Dock and being briefed on practice D-Day landings taking place on a beach down by St. Anne's Head? Those on their rotational day off were told so they could draw 22's and shotguns, go down to the "invasion", hide there, and shoot rabbits and quail scared out of their nests by the tanks and half-ton trucks coming ashore. Boy! Sometimes those guys in the tanks thought they were in a real invasion when our sharpshooters fired away. All this to supplement our rations that the British were so good to share with us.

* * * * *

BEAUTIFUL TEXAS: The time while flying from the Salton Sea to Corpus Christi passing the Texas border and Bill Craddock yelling about the green pastures when there was nothing as far as the eye could see but burned out sage brush and dried up creeks?

* * * * *

CHATTER: Sniffer never letting a mouse get away in a prescribed area? The great "all you could eat" shrimp and wild boar cookouts - lasted for two days in back of the Quonset huts at Port Lyautey? The average loss of weight of 33 pounds per man in the Squadron after six months at Pembroke Dock? The knowledge that Rod Brush and I

The Story of VP-63

might have influenced Jim Hardy about that girl he had in San Francisco – now his gracious and lovely wife, Jean?

* * * * *

FRENCH BREAD: Remember how great it was to return to Base at Port Lyautey in the 3am French bread 6x6, all nice and warm and smelling so good (often missing the midnight bus back to Base)?

The unbelievable goodness of fresh French bread and 2-pound squares of butter to greet us at Port Lyautey after our long stay at Pembroke Dock? (Ed's Note: Don't forget about the dozen eggs some of us ate the first breakfast in Port Lyautey, Fairlie.)

* * * * *

Fairlie says, "Looking forward to the reunion which must be 'impossible' because it took a little longer." I occasioned to visit Pembroke Dock a year ago. Texaco now has a pier extending through our take-off and landing area, but would you believe the old single hangar at the top of the single ramp is still there but in disrepair? The original fence around the plane ramp is still there, but has not seen any use for years. The Officer's Mess and Briefing buildings remain, but the EM building s and mess are gone and replaced by a shipping line. Texaco has taken over the practice bombing lagoon west of the base and oil storage tanks are all over the place. But, the town and Tenby do not seem to have changed at all. I suffered from nostalgia while I was there. I was a pilot for Pan Am from 1945 - 1980 and am now retired in Tequesta, Florida, and running a charter trawler for a week to ten days in the Bahamas.

* * * * *

MAD Cats

Lt. (jg) C.L. Knight

ANYONE FOR A SIGHTSEEING TOUR? While returning from patrol in the Bay of Biscay, we received a weather report that everything was closed in. Doc Spears proceeded around the coast and, with a break in the weather, managed to make an open sea landing at the Isle of Scilly on the southwestern tip of England. The plane was taxied inside a breakwater and anchored. The crew spent the night aboard the plane; Doc and I stayed on a very small patrol boat. The next day, the weather cleared and, after calling on the local dignitaries, Doc proceeded to take several for a flight around the island.

* * * * *

BOMB THE BISMARCK? While on the return leg of the patrol, Crew 10 spotted a large oil slick. Since we had been alerted to be on the lookout for a crippled sub, we were sure that our luck had changed. When we flew over the oil slick with MAD on, the signal received was definitely not that of a sub. A message was sent to base reporting the oil slick and the contact as being doubtful sub contact. The British came back. WAIT! Finally we requested permission to bomb. They came back with, "Negative; continue to patrol." When we returned from patrol, we were advised that they didn't want us to bomb the contact because it was the sunken Battleship Bismarck, and its location was too close to a large mine field.

* * * * *

SMALL BOAT TO STARBOARD: While operating out of Gibraltar (I wasn't with Doc on this trip), I believe Simer and an ensign by the name of McDougal were flying with him. I believe McDougal was in the right co-pilot seat on take-off. Doc was at the controls and McDougal failed to advise him of a small boat pulling away from a British ship and crossing the plane's path from starboard to port. When Doc did finally see it, it was already too late. He hoisted the

The Story of VP-63

plane up; however, the starboard float struck the cabin of the small craft, knocking the float completely off. Nobody was injured; however, McDougal was assigned to remain in Gibraltar until the float was replaced.

* * * * *

DUCK HUNTING PBY? Soon after joining the Squadron in Alameda, I was assigned to Doc's crew. While on training flights in lower San Francisco Bay, the PBY proved to be an excellent weapon for hunting ducks. (Ed's Note: Now hear this, Skipper!)

H.D. Lee, CAP

BAD GAS? I remember one dark night about 800 miles out of Bathhurst en route to Natal when both engines started to cut out at the same time. I never did find out what caused it and it happened several more times before we got back to the States. A bigger thrill you'll never have. (Ed's Note: Maybe you were using GASOALKYHOL, Lee.)

* * * * *

SEAT OF HIS PANTS LANDING: I remember the night we took off from Angel Bay with a full load of gas, passengers, baggage, and a dog or two. We were heading to Port Lyautey. In very short order, we discovered that we did not have any of our pressure instruments, no air speed, no altimeter, or rate of climb indicators working. Virgil Mayabb was flying and I was co-pilot. (It just dawned on me that the bad things seemed to happen when we were up front.) There was no other alternative but to land, and landing in Angle Bay at night was hairy enough with everything working. The landing was accomplished using lots of power and trying to keep a nose high attitude. The culprit turned out to be wax in the pitot tube which was caused by the pitot heat being left on during the taxiing from Pembroke Dock to Angle Bay that morning.

* * * * *

FLARE GUN: I can recall one morning after leaving the buoy at Pembroke Dock and we were taxiing to the takeoff area when suddenly there was the sound of a lot of activity coming from the after-station. It seems that George Ackerman, our ordinanceman, was investigating the complaint that one of our flare guns would not close properly when loaded with one of the long shells. The gun accidently fired, hitting Ackerman in the palm of the hand. Crewmen used one whole bottle of CO_2 fire bottle on the flare which was burning furiously in the bilges. Had we not been on the water, which helped to dissipate the extreme heat, it probably would have burned a hole in the bottom of the plane. We immediately radioed for a dinghy to take George ashore for medical attention. He spent some time in a British hospital and rejoined our crew at Port Lyautey.

* * * * *

BUOY SNATCHER: I remember a big strong radioman (probably Bobby Henderson) who had a habit of catching a buoy in one hand, while holding the bow turret with the other, would stop the plane. This is one way to tie up a PBY, but it's not recommended. This young man had been successful in pulling this off several times, much to the amazement of his fellow crewmen. His luck failed him, though, being a bit surprised at this, made a quick decision and let go of the plane instead of the buoy. By the time that plane could make another pass and pick him up, he had become a very cold sailor.

The Story of VP-63

(US Navy)
CDR Robert A. Barton gets a last minute update from Lt. Sam R. Parker before takeoff

Lt. (jg) W.P. Craddock

THE LONGEST DAY: It was August 2, 1943, the day after Lt. Tanner and his crew flying the *Catalina* affectionately called *Aunt Minnie* had been shot down by German fighter aircraft in the Bay of Biscay. The area of the Bay where this occurred was known as "Junkers Junction", the nearest point to the French Coast where the Germans had based their fighter aircraft, JU-88s, to protect incoming submarines from Atlantic patrols. Lt. Jim Hardy PPC of one of the two planes from our squadron assigned to search for the survivors from the *Aunt Minnie*. Lt. (jg) Fairlie and I took turns flying in the co-pilot seat and navigating for 16 hours on an east-west course which terminated on the west end of "Junkers Junction" and on the east side about 10 miles from landfall on the French Coast.

August 2nd was a day no aviator in VP-63 cherished to fly in the Bay. It was beautiful and cloudless; one with only gentle breezes and a blue, blue sky that went on to the end of the world. Why we didn't run into

MAD Cats

enemy fighters that day is beyond my wildest imagination. They should have known we would be out looking for any survivors. It was either great luck or a German holiday!

As Jim Hardy sat there in the driver's seat, he seemed quite and unconcerned of the dangerous situation we were in. We were really sitting ducks for the Germans! I believe in retrospect, and after regarding his stories that appear in this volume, that he was too busy praying to be otherwise. On our leg in toward France he would just sit there, kinda like he was asleep but his eyes were open, and keep on going east closer and closer to the French coast. I thought any minute we would be over Brest or Paris or wherever. It seemed an eternity of stress before he would turn around and head west again. It was a relief to be heading west although it would have been easy for the fighters to overtake us all the way out to "Junkers Junction" and beyond.

Back and forth we flew all day in this environment. After flying over water this long looking and searching for aircraft, submarines, and people in the water, you begin to see all kinds of mirages - periscopes, ships, aircraft, elephants, yellow submarines, guille birds, etc. I bet I saw ten periscopes and hundreds of planes that day, but no sooner seen than they would disappear. Lunch didn't taste very good as it never did when the going got rough in the Bay. You know how hard it is to swallow when you haven't any saliva. That's the way it was.

Finally, in close to the French Coast, we spotted a rubber raft in the water, but empty. At the time, we didn't know that there had been any survivors from the *Aunt Minnie* because British intelligence had intercepted a German message the day before saying that there ware no survivors. At any rate, we circled the raft for about 15 minutes to be sure. Satisfied there was no one in the water (at this time we did not know that a ship had already picked up three survivors from the raft), we headed home as darkness began to overtake us. What a relief to see darkness coming on and to be heading home again. The prospects of kidney pie, Brussels sprouts, and marmalade again almost brought the much missing saliva back to my mouth. This ended the

The Story of VP-63

LONGEST DAY for me, although there were a lot of long ones in the Bay.

* * * * *

ABANDON SHIP! ABANDON SHIP! It was a bitterly cold morning on the 2nd of December, 1943, when Lt. Knorr and I took off on a routine patrol of the Bay of Biscay. It was a quiet morning, little wind blowing, which made it possible to take off to the west. Usually the prevailing wind was from the west, forcing long taxis to the west bay for takeoffs into the wind.

As usual, two planes made the daily patrols, one flying the inner leg and the other, the outer one. On this morning, both planes had entered their position for takeoff at about the same time after routinely checking the engines. The other plane was to the north side of us.

As it would happen, both planes took off at the same time, but on a converging course. As we got on the 'step', the other plane kept getting closer and closer to us. I kept signaling Knorr, the PPC for the flight, to move to port to avoid a collision. He kept kicking left rudder and we moved gradually in that direction and into a more comfortable position with respect to the other plane's flight path. However, we did not pay a lot of attention, at the moment, to a fishing trawler positioned ahead and to port making ready for sea. There were four seaman busy in the deck making ready, and it was only a microsecond later (to use Nick's terminology) that we got their undivided attention. Our port wing came so close to the side of that ship that I thought we only missed it by an inch or so. Just after we cleared the water and had cut back on the throttles, the man in the after-station called the PPC on the intercom and said, "Sir, those men on that trawler just jumped into the water!" It was cold enough for them to break ice when they slammed in.

On our return to base from patrol, we went over and looked at the port wing and found that the thin metal overlay piece at the end had been bent flush against the butt end of the wing. That was a narrow escape from disaster; but if those seamen had been able to identify us later, we would have had to deal with another type of disaster, I'm afraid. *C'est le Guerre*! ABANDON SHIP? AYE, AYE!

* * * * *

SKY ROCKETS IN FLIGHT - NOT AN EVENING OF DELIGHT!
If you've never experienced shore anti-aircraft batteries firing at you at night, then you've never lived!

We used to come into Pembroke Dock at night after a long and weary flight and have those British rascals open fire on us. In terrible weather most of the day, and lost a good part of the time, and finally getting our position for final approach into the base, these guys would begin their exercise with vim and vigor.

When you first observe the tracers, fed into the ammunition belt about one in every three cartridges, they appear to be moving to some far away object. However, it's amazing how these bullets start out in a line about parallel to your course, but apparently some distance away from your plane they then curve in a big arc right at you. Each tracer appears as though it will hit you right between the eyes, and many times I have ducked while sitting in the co-pilot's seat.

I never did find out whether those shore battery guys were just practicing with no intent of hitting us or were dead earnest. Seeing how the German Air Force bombed out Pembroke Dock before our arrival didn't help our sense of security knowing that they were probably super sensitive and trigger happy when it came to airplanes flying around their bailiwick. SKYROCKETS IN FLIGHT - NOT AN EVENING OF DELIGHT!

* * * * *

The Story of VP-63

THE INVISIBLE CITY: I wonder if anyone else ever experienced seeing a whole town of lights, 200 miles from shore, and being unable to spot anything at all on the water. We've flown over these lights several times with radar on and have yet to identify any objects on the surface. Really spooky! If it had been a fleet of fishing vessels, we should have been able to pick them up on radar - but nothing! Maybe phosphorescent fish?

* * * * *

A FLYING SUBMARINE: On September 6, 1943 we were flying a routine mission in the Bay. About one hour after departing "Junkers Junction", moving on our south course, I was sitting in the after-station eating a snack and noticed a B-24 bomber about three miles to port. We were flying low, under 1,000 feet at the time, and the B-24 was judged to be at an altitude of 3,000 and 4,000 feet. Lt. Hardy was the PPC and Lt. (jg) Fairlie was the co-pilot. I was the navigator.

I watched the B-24 for at least a half hour and he was on a course which would eventually take him across our flight path. When he finally crossed in front of us he was still at least 2,000 feet above us. When only a short distance ahead, he suddenly turned very sharply to port and came directly back toward us. While still not completely out of his sharp turn, he released depth charges which hit the water about 200 yards away from us. I immediately called Hardy on the intercom and directed him to the point of contact of the bombs with the water. Descending rapidly to 50 feet above the ocean surface, we dropped a float light, set the MAD gear, and started our spiral search procedure. At each 90 degrees, another smoke light was dropped.

By the time we had made two complete circles, the B-24 joined up and was following us around the spiral, but at a much safer altitude, perhaps 500 feet. Apparently the B-24 radioman had sent a submarine sight signal, a routine 465-472 code, which soon brought four *Sunderlands* to the scene. All planes joined up behind us in a stepped-up echelon formation with the four *Sunderlands* between 500 and

MAD Cats

1,000 feet. All were following around and around as we progressed on our spiral procedure, waiting for further orders for us, since we were in command as the first plane to enter the fray. What a beautiful sight this would have been for a swarm of Ju-88 fighter planes. After flying our pattern for about half an hour, we broke off the charade and continued on our pre-assigned mission. All other planes went back on their routine patrols. No MAD contacts were registered on our equipment.

After 14 hours in the Bay, we returned to our home base. As we approached the beach in the dinghy, and after securing our plane to a buoy, we noticed quite a contingency of brass waiting for us. Included in the group were the Base Commandant, his Executive Officer, our Skipper and Executive Officer, a Communications Officer, and a Duty Officer. Quite a welcoming party, to say the least, for a routine patrol. We all thought, "What have we done now?"

As it turned out, this is the way the events went regarding the B-24 experience. When the B-24 crossed our bow, the tail gunner shouted "Submarine!" in his mike. With this, the B-24 pilot made a sharp turn and released his depth charges before completely finishing his turn. He must have squeezed his trigger-happy finger when he saw the first blur. In the meantime, his radioman sent out his "submarine-sighted" signal. When the B-24 crew realized their mistake, and obviously thinking the PBY was on the water, they signaled that they had bombed a PBY on the water.

The last message was giving gray hairs to some people back at the base. The poor radioman in the B-24 didn't know what the hell was going on since he couldn't see anything, so every time someone said or shouted something, he transmitted it over the radio. Correcting the last message with another, it read: "Have not bombed a PBY on the water; have bombed a PBY in the air!" This really made people wonder back at the base. Well, some way or another, they found out we were alright and they quit worrying, but they were still mighty confused. All of this was happening unbeknownst to us. We thought they had

The Story of VP-63

really sighted a submarine below us and we never in this world imagined we were the target.

Well, maybe we did look like a submarine, and perhaps this is the reason we didn't get jumped by their fighter aircraft anymore than we did. After all, what good German is going to go around shooting up one of his submarines, even if they are airborne. Flying over all that water for a long time will do strange things to people, maybe even the "Super Race Guys" and the Army Air Corps. A FLYING SUBMARINE? BAH!

* * * * *

THE GERMANS ARE COMING! THE GERMANS ARE COMING! In the latter part of September, 1943, we were flying back to Hamworthy, an English alternate port, when we couldn't get back to our base at Pembroke Dock due to inclement weather. About an hour from Pembroke I, as navigator, gave a routine message to the radioman giving the base our estimate time of arrival (ETA). About 15 minutes after the message was sent, I noticed an unusual amount of British air activity past and around us. Fighters hurried past on their way seaward. We didn't think much of it at the time, but later, on our arrival at the base, it became clear why all the activity.

Again, as it had happened before, we were greeted, upon arriving at the beach, by a distinguished gathering of high-ranking officers. The difference was that they had a much more tense expression on their faces this time. They immediately wanted to know what the devil was going on. This was a difficult question to answer since we had just returned from a very routine return-to-base-from-an-alternate-port flight.

It didn't take us long to find out what happened. Since we didn't have the present day code book, we were forced to use the only one we had yesterday's! The first code in our message was a standard code used by everyone to notify the base we were using the prior day's code book and to refer to that one to decipher the message. When the British

MAD Cats

received the message and tried to decipher it with the current day's book, the message made no sense at all. They immediately thought: "Those crazy Americans will do anything,; they've sent a message in English and disregarded orders to always use codes." Then their imaginations went wild when the message appeared to spell out a major German invasion force approaching Lands End on the southeast coast of England. The coded message was as follows: 1BAS, 5DSD, 3CRS, 50BMS, 16SVS. The British knew our position from a radio fix when we used the radio and they interpreted the message as follows: 1 battleship, 5 destroyers, 3 cruisers, 50 bombers, and 16 support vessels just off the coast of England poised for a major invasion of the British Isles. The alarm went out all over England and the country immediately came under alert. No wonder those stern and tense faces that greeted us at the dock; the top British admiral had them on the phone constantly until our arrival wanting to know what the bloody hell was going on up there. The message scared him out of his wits. Of course, no trace of the invasion force was detected by half the RAF.

I had to talk to the top admiral on the phone from the Communications Center. Why not? He had talked to everyone else in England by that time. I knew I'd better have an awfully good explanation. Well, in my Texas brogue, which, by the way, was worse than his, I attempted to give a reasonable summary of events that had ensued unbeknownst to me all the while. He apparently accepted my cursory review, rather reluctantly, kinda like he didn't believe such a thing could happen in HIS war. Inwardly, I felt the same way about it. I always wondered why this sort of thing hadn't happened before. Maybe it had, for all I know. But, I think in this case everything had to work out just right or wrong sequence, and with the right or wrong personnel for it to come off the way it did. It was probably just a matter of timing - being in the right place at the wrong time.

I bet the British Communications Officer had a red face when he realized he had taken too much for granted. Maybe, for once, he didn't think the Americans were the only crazy people on earth. At any rate, I was sure glad I was right, because I might have been severely punished and sent home. I couldn't have accepted this sentence very gracefully knowing full well what fun I would have missed by not being with all the fellows in the Bay and French Morocco. THE GERMANS ARE COMING! BAH!

PBY Radio Compartment. Source: U.S. Navy

Lt. H.J. Baker

VP-63 sank U-761 (*Geider*) - February 24, 1944, Strait of Gibraltar

63-P-14   Lt. H.J. Baker
Lt. (jg) R. Brush
Ens. W. McSharry
E. Tanneberg, AMM
W. Franklin, AMM
P. Pearson, AMM
M. Crider, ARM
C. Gravel, ARM
J. Jellison, AOM

63-P-15   Lt. T.R. Woolley
Lt. (jg) E.W. Kellogg
Ens. R.D.J. McCarty
R.H. Whalen, AP1c
J. Cunningham, ARM
L. Coker, AOM
Bob Henderson, ARM
B.F. Martin, AMM
D.I. Mayhew, AMM
M.B. Cummins, AMM1c

1559	Planes gained contact on submerged U-Boat; tracked with retro-float lights.
1610	HMS *Anthony* entered pattern, contact lost.
1615	Set up expanding spiral (2 planes); two British frigates cleared area.
1645	Contact regained – tracked with lights.
1656-1658	63-P-15 and 63-P-14 attacked with 24 bombs each.
1702	U-761 surfaced out of control (bow only).
1704	U-761 surfaced out of control.
1707	HMS *Anthony* and HMS *Wiseheart* dropped depth charges.
1710-1720	U-761 resurfaced stern down; commenced abandoning ship and set off scuttling charges.
1717	PV from VB-127 arrived from patrol 20 miles to the west and dropped depth charges.
1718	British PBY circled scene.
1720	U-761 sank; 48 survivors including CO; picked up by frigates and taken to Gibraltar.

The Story of VP-63

U-761 FOLLOW-UP: VP-63 PBYs were ordered to remain on patrol until relieved; hence, arrived at Port Lyautey some time after 127 PV and then were on buoys for some time during a rainstorm and boating delay.

In the meantime, the 127 crew and 'brass' were the guests of the Commodore in his mess and they were claiming full credit for the sub sinking. Only after Lt. Woolley and I were debriefed did the full story come out -- that they had depth charged a crew abandoning ship from a sinking sub. As I recall, they (127) were ushered out and replaced by the VP-63 skipper and the crews of 63-P-14 and 63-P-15. The VB 127 crew were awarded DFCs and AMs within two weeks at Port Lyautey. Lt. Baker and crew waited over a year (3 April 1945) for their awards at Dunkeswell, England.

Result of MAD Cat attack. Source: VP-63

MAD Cats

(U.S. Navy photo)
Crew Chief Earl Taebert and LT Howard Baker beam at U-boat silhouette on their *Catalina*

THE FAMILY JEWELS: The Hardy Boys on TV have nothing on the Hardy Boys of VP-63. Only 45 Greek submariners, an English Liaison Officer, and Lt. Jeff Baker remember when Lt. Jim Hardy nearly left the family jewels draped on the periscope combing of a Greek sub. He was clearing the Conning Tower on dive 15 of 19 crash dives in a 3-hour period while acting as a target observer during simulated attacks from VP-63 planes out of Port Lyautey. His comments were not part of his official report.

* * * * *

McSHARRY TURN WHITE? Who can forget Ens. Bill McSharry's renderings of *The Great Speckled Bird*? It will probably be a surprise to all his buddies to know he earned his living as a San Francisco Building Construction Inspector, and not as a driller of Keesters in sawhorses (as he told everyone he was going to). A bit ruddy of complexion (to say the least), he was practically unrecognizable when a shore battery hit on 63-P-14 turned him pale for the only time in his career. A real ship-mate in the truest sense of the word, his crew mates could always count on him for a chuckle. We all remember a typically nasty English night while circling Milford Haven on instruments, waiting for landing clearance, and having determined the hard way that the merchant fleet at anchor had not hauled in all their barrage balloons. Bill was asked over the intercom if he wanted to take over in the cockpit, and he replied, "Sorry, but I'm going to stay right here in the waist hatch in my rubber boat, with oars ready, until you get this damn tub down."

* * * * *

FATE WORSE THAN DEATH DEPARTMENT: Never wish it on anyone to be promoted (demoted?) from Warrant Gunner to Ensign while enroute to a new squadron assignment. Being the Bull Ensign in a squadron usually has its perks; but being the only Ensign at Pembroke Dock bordered on the ridiculous – going from only Ensign to Senior JG due to a backdated promotion.

* * * * *

BAR BILL – RAF PEMBROKE DOCK: Lt. Benscoter usually led the pack with his roommate not far behind. The outstanding memory, though, is how good a full glass of neat Scotch tasted after a 24-hour day involving a Bay of Biscay patrol, especially when your roommate was buying.

* * * * *

SMOOTH TRANSITION: C.H. Hutch's shift from Shove Ha' Penny aficionado to Barn Yard Golf King of Lyautey. (Hoss Shoes!)

MAD Cats

BAHIA BELLE PROMENADE: Sounds like a fun craft, but doubt it will match the really appreciated rides from Milford Haven to Pembroke Dock after a long day in the Bay of Biscay.

* * * * *

DECEMBER 7, 1941: Jeff Baker was in Panama at Coco Solo [NAS Upham Field, AOM1c, NAP] attached to VP-32 flying PBY-3s. December 7, 1941, marked the commencement of many months of long patrols in the Pacific, eight days out of Panama – with ROWs swinging on buoys in the Gulf of Fonseca, Galapagos Islands, and off the coast of Ecuador; refueling from tenders and converted merchant oilers. We weren't exactly prepared for December 7. I recall a Gas Attack Alarm one evening, when in the absence of masks, all hands were instructed to walk into the bay with wet blankets wrapped around their heads. As I recall, the vast majority decided to take their chances with the yet-to-come gas attack. Looking back, tension was at all levels of command, as no one knew for some time where the (Japanese) fleet was, and the Panama Canal was high on the anticipated list of follow-up attacks. On another December 7th, but 1944, Jeff and Vicki were married. (Ed's Note: Are you trying to tell us something, Jeff?)

Lt. (jg) C.R. Rehn

WOOL SHIRT: I have one item of memorabilia which reminds me of our short stay in Iceland. It is my black wool Navy shirt which was issued when we arrived at Reykjavik...even if it was July. The shirt and my Blue Nose Certificate are assets to have to survive the sub-zero winters in Minnesota. I keep thinking that when the shirt wears out I will have to move to Arizona or California or Texas, but it looks like it will last another 20 years or so. I remember what a surprise the "Land of the Midnight Sun" was - I expected pine trees and lakes, not a rock pile with underground hot springs! However, it

The Story of VP-63

did prepare me for a similar looking area when I was out on Adak, Alaska during the Korean War.

(Photo: VP-63)
Reykjavik, Iceland

* * * * *

BEN'S CIGAR: I was one of the very few non-smokers to come out of WWII. I must give full credit to PPC Benscoter and Plane Captain Woods. On one of our 15-hour patrols down in the Bay Biscay, the cockpit was filling with cigar smoke. Ben suggested that I try smoking one of his cigars because it might help us spot a U-Boat. After puffing away for a time, Woods brought me a nice cup of coffee...with curdled cream. The combination was too much and I spent the rest of the flight in the after-station. It permanently killed any desire I might have had to smoke! During those long patrols in August '43, I spent a good share of the time navigating. The code OA (Under Enemy Aircraft Attack) still sticks in my mind. Fortunately our plane was never intercepted. I always thought it was the camouflage created by Ben's cigar smoke that kept us well concealed!

* * * * *

QUEEN'S SUITE? I remember an R&R trip to London as an experience of questionable hospitality by our British allies. I was

MAD Cats

given the Queen's Suite on the top floor of the famous Dorchester Hotel across from Hyde Park. I thought it was really considerate of the British to provide all that luxury (even gold-plated faucets) for a U.S. Navy Lieutenant (jg)! That night I quickly discovered why this suite was so readily available. When the buzz bombs started exploding, I caught on a hurry that the most desirable rooms were on the ground floor – or even below!

* * * * *

BULLY GOOD SEAMANSHIP: There was a time the British Harbormaster greeted me after I tied a PBY to the buoy between two *Sunderlands*. This was my first assignment as a qualified PPC. My instructions were to move the PBY from the outer harbor to the inner one. I was given a pickup crew. Everything was going fine and we were just about to hook onto a buoy. Using the intercom, I told the Tower Plane Captain to "cut engines". Instead, the floats raised! Fortunately the tide and the wind were just about right and we made a complete 180 degree turn around the buoy. The wings were righted, the floats lowered, and on the next pass we made it right between the two *Sunderlands*. The Harbor Master met me with, "Good show! Bully good seamanship!!" Little did he know how I had sweated during that maneuver. It could have been the end of three flying boats and my Navy career!

WILD BOAR FEAST: Another standout in my memory was the Wild Boar feast at Port Lyautey. It was an all-day barbecue supervised by the Chief Flight Surgeon from Texas. The sauce was hot and spicy, Texas style, and we had lots of beer to cool us down. It was at this event that I met another Navy Pilot who was with VP-73, the squadron we were relieving. He was also from Minneapolis and often said he had never seen a squadron quite like VP-63! He became a good friend and was influential in my staying in a Navy Reserve squadron in Minneapolis. In July 1950, this squadron was recalled during the Korean War and we flew anti-submarine patrols out of the Naval Air Station at Kodiak, Alaska.

The Story of VP-63

* * * * *

<u>MAY 29, 1945</u>: A flight that really stands out in my memory was on May 29, 1945, a flight to observe the English Channel after VE Day. There were 14 passengers on board. It was a sight I'll never forget. The ships were all running a normal course and not a zigzagging pattern. The coastal towns were all lit up with lights blazing and all the flags flying. I felt that our efforts and the long patrols had contributed greatly to the end of the war and this peaceful scene. I still have a great sense of pride in our VP-63 Squadron and the efforts that helped achieve the final victory.

* * * * *

<u>DOG SQUADRON</u>: I remember coming back to the States after VE Day and landing in Jacksonville. How could anyone forget the looks on the faces of the "welcoming committee" when all the dogs that had been acquired in England jumped out of the planes!

MAD Cats

R.I. Weinrub, ARM

<u>ALL SWITCHES OFF</u>: It was Saturday, July 22, 1944 at Port Lyautey. Second Radioman Bonanno shook me and said, "Come on, Ray, it's time to move out." It was 0430 on a clear, muggy morning with just a trace of light beginning to show. I threw off the mosquito netting, climbed out of the wooden, folding cot, and mumbled something about the ungodly hour. We had been in North Africa for five months and it was impossible to get used to the heat and 100% humidity. The mosquitos were as big as my thumb and ate me alive, and the air was full of pungent odors from the Arab Medina a couple of miles from the base.

This particular day started out the same as any other when we were going on patrol in the Strait of Gibraltar - hunting U-boats at 100 feet off the water. We took off from the river at about 0600 and headed for the Strait about 150 miles away. Lt. Vopatek was the PPC, navigator. I believe Gaskin was the Chief Mech, and names like Kinkle, Gross, and Keane seem to come to mind, but I'm not sure.

About 1300, right after lunch, I was standing behind Bonanno when I smelled gas. I told Vopatek and then turned back toward the radio compartment when all hell broke loose. Gas was pouring in over the center of the bulkhead above all the electrical area. I leaned over Bonanno and hit every switch possible until all were off. While this was going on we had climbed to about 2,000 feet, but now, without power, we started down. Gaskin, or whoever, was in the tower had to crank the floats down by hand, as I had cut everything. I figured one spark and we were gone!

It doesn't take long to go from 2,000 feet to zero when you have no power. Lt. Vopatek made a beautiful full-stall landing, but we hit a wave at just the right angle and went shooting into the air about 160 feet. Then - straight down - POW!!

I was thrown against the pilot compartment bulkhead and was dazed for a few minutes. When I started to get up, there was a propeller

The Story of VP-63

about a foot from me on each side. We had hit so hard that we dropped both engines behind the oil cells, and the props had cut right through the aircraft. Water was starting to come in and confusion was everywhere. The co-pilot (Gillespie?) had his scalp cut badly and Lt. Vopatek was trying to help him. I headed for the aft end to get the hell out when Gaskin made me go back and destroy the MAD gear. In the meantime, Socha, or someone, had put the 7-man raft over the side, but didn't tie it to the plane and it floated off. Thus we had only a three-man raft and Lt. Vopatek and Gillespie and one other got in the raft and managed to stop the bleeding head.

As water was coming in, the rest of us jumped onto the water and hung onto the side of the small craft. The water wasn't cold and I seem to remember a brandy being passed around.

At about that time, a B-24 from Oman came over and spotted us. They signaled that they would report our plight to Algeria who, in turn, would contact our base. They told us that help was on the way and headed away.

I can't remember exactly, but I believe it was 5 or 6 hours before we saw the rescue craft with the American flag flowing in the breeze. What a sight!! I don't remember what we all talked about except that we were all lucky to be alive. The only strange thing was that the plane was still afloat the next day and they had to go out and bomb it to sink it.

The incident occurred about 39 years ago, but some of it seems like yesterday.

There was a Naval inquiry later and they were very disturbed about my cutting switches so quickly; however, on October 15, 1944, one of our planes blew up at sea and I never heard any more!!

MAD Cats

P.F. Ware, AMM

<u>OF GAS LEAKS</u>: Years ago I related the incident about the "gas leak" a few times, but people listened with skepticism and doubt, so I had all but forgotten it and assumed it to be a figment of my imagination.

As Lt. Craddock will recall, it was on our flight back to Africa with our old plane fresh out of overhaul at Norfolk A and R. According to my Navy flight log, it was 0230/12-22-44 in Plane #8 between Natal, S.A. and Bathhurst, N.A. The log shows that Lt. Craddock was PPC and I was Plane Captain. Ens. Heath was co-pilot; Ens. Bowen, the navigator; and the crew was composed of Gaskin, Scarsella, Chaisson, Comstock, and Boreland.

I was half asleep in the tower when Ens. Bowen announced over the intercom that gas was leaking through the overhead. One look and I cut the main switch which turned off all the lights, including the pilot's instrument panel. We had to scramble for a flashlight for the pilot's use. Something had to be done quickly or we would become a fireball. After a hurried consultation with Lt. Craddock, it was decided to try to have a look in the cabane section where all the gas valves and connections were located. Lt. Craddock must have advised me to do this, as I don't think I would have volunteered to do anything so foolish. (Ed's Note: As I recall, Ware did this voluntarily without telling the PPC. The PPC would not have thought of it as he didn't know it was possible to reach this section from inside.) So, with a gun belt around my waist and lashed to the navigation table with Bowen and Chaisson holding my ankles, we opened the navigation hatch. Lt. Craddock had dropped the airspeed down so the slip-stream was not as great as we had anticipated.

Standing on the navigation table, I could reach the port inspection panel over the main gas valves and strainer area. After unlocking about three zeus fasteners, the slip-stream ripped the entire panel off, thus breaking out the port blister plexiglass. Working with only one hand and by feel, I tightened the gas line clamps that I could reach.

The Story of VP-63

This couldn't have taken more than five minutes, but it seemed like a five hours! After dropping back inside, Bowen was still mopping the overhead with a rag and advised us that it appeared the leak had stopped. The plane was opened up, fore and aft, to air it out. By this time, it smelled like a gas station.

The remaining portion of the flight was uneventful except for the oil leak in the port engine that pumped about 30 gallons of oil into the deicer system and, thus, into the port wing. All of this did serve to get the adrenalin flowing and nobody was the least bit sleepy for the rest of the night. (Ed's Note: The gas leak was so bad that I'm convinced that Ware's action saved the flight from becoming a fireball, or at best, a forced landing at night in a rough sea – either of which could have written finis to us all.)

* * * * *

MOVIE STAR BOWEN: Were you aware that Ens. Bowen was in two post-war movies while attached to the Kearsarge? In the movie **Bridges at Toko-Ri**, he was Catapault Officer and was shown sending jets off the catapault. In **Caine Mutiny**, he was the O.D. as Fred McMurray was piped aboard to see the Admiral.

* * * * *

THOSE COLD WINTER EVENINGS: On a cold winter evening as you casually turn your thermostat up to warm your home, do you ever think back to those cold and damp nights in England when we were trying to keep warm with one fireplace in the center of our quarters and no fuel? Recall that we took turns stealing a bucket full of coke from the power plant next door until some luckless person was caught. It was then that the teakwood paving blocks between the old submarine slipways began to disappear as they burned better than coke. That, too, came to a sudden halt. Some ingenious soul then rigged a 5 gallon can with the appropriate tubing and valve system to

MAD Cats

drip used engine oil into a sand bucket. A bit of a fire hazard, but it did have a warming effect.

* * * * *

PERFECTION: Remember Lt. 'Doc' Spears always striving for perfection, even to practicing his bombing technique by making a parachute from a handkerchief and, weighted with 50 caliber shell casing and a note, dropping it over the residential area of Pembroke Dock on return from patrol in the Bay?

MAD Cat Retrobomb attack. Smoke floats for tracking can be seen in the background. Source: VP-63

Lt. (jg) J.C. Fox

OF ICY LAKES: In January, 1945, Lt. Carl Benscoter headed up a detachment of four planes, including six crews, dispatched to Dunkeswell, England, from Port Lyautey. I was in a plane PPC'd by Lt. Bill Ray, and flew, the first day, to Marseille, France for a RON (Remain Over Night). We filed with the Army Air Corps and they gave us our navigation data, including some strip maps something like the AAA would give for a car trip.

The Story of VP-63

We formed a loose formation upon leaving Marseille and saw the last of each other when we turned north at Toulouse. We went into weather, and it was cold as hell. From that point it was sort of dealer's choice. I don't know what the others did, but Lt. Ray decided to climb to 14,000 feet! We stayed up there for about an hour and then decided to descend. We broke out around 1,500 feet without the slightest idea where we were except over land. There was no navigation aide for a course check, only the Paris A/N range west leg for a speed/altitude check, and they must have been about 16 miles wide as far west as we were. It didn't matter - we never heard it anyway.

Being more or less lost, Lt. Ray resorted to an Air Corps tactic. He encountered a well-used railroad track and along it we went for a half hour or so. Nothing to be seen. Reversing course, we flew the same track for about an hour. No joy. Then, as I recall, we turned west, perhaps on the theory that the Atlantic Ocean was west of France. We found it and a large town of unknown identification at the time. There was an ample bay, but it was decided not to land, which was fortunate since the town was Lorient and it was German occupied. They did not shoot at us, there being some sort of truce in force (we later learned).

Then Lt. Ray flew north, perhaps on the theory that England was north of France and separated by the English Channel. Sure enough, we found the English Channel and also an airfield. This field had been demolished by the Germans, but the French had filled in all the holes. This fact not withstanding, Lt. Ray elected not to attempt a landing. Instead, we headed in what we thought was the general direction of a French town called Rennes on the theory that we could get some directions by radio from the British who were supposedly based there. Meanwhile, time passed and the sun began to set. Then it was noted that below us lay a body of water suitable for landing a seaplane of our proportions and I, now sitting in the co-pilot seat, suggested that we land in it. The body of water was a damned reservoir and very deep except along the shore line.

We landed and anchored. We had attracted the attention of the people of a small village later to be determined to be Glomel. Soon a boat

MAD Cats

and two Frenchmen appeared. After some halting exchanges, we determined we were in Brittany and I was sent ashore to negotiate for places to bed down for the night. Suffice it to say, we were treated royally after proving ourselves to be Americans and not the Luftwaffe (both wore green uniforms). We gave them oranges and they gave us calvados.

The next morning we discovered about 6 inches of snow had fallen. The weather was not quite VFR. The airplane had to be cleaned off, so more of us suggested that tomorrow was another day and the calvados were in plentiful supply. These mild protestations were met with the ultimatum that either we should get aboard or seek alternate means of transportation. We got aboard. We cleaned off the wings and tail. I was now the navigator, using an Esso map of Europe where distance to Rennes (where we were now expected after phone calls) was roughly 3/8 of an inch.

Picture a finger and thumb with the finger two miles long and the thumb about one mile long. You are at the base of the thumb and if you choose the right course, you head across the palm and toward the fingertip. If you proceed up the thumb, so to speak, there is less distance. At the base of the thumb is the reservoir dam. It was cold! When ready for takeoff, away we went and I believe everyone expected we would take the two-mile takeoff course. Instead, we took the short course and, no doubt, there must have been a good reason for it. Takeoff seemed long and slow, perhaps because the thumb was slightly frozen over, thereby impeding progress to some degree.

Suffice it to say that we just plain ran out of water. Lt. Ray horsed it in the air and there was general relief for a moment. This relief was abruptly interrupted by a strange sound barely heard by seaplaneists. It was the sound of a giant oil can scraping the earth created by the contact of the seaplane keel with the hard, frozen ground. Before the final crash, it is assumed that the throttles were closed. Our forward momentum, nevertheless, took us through Hedgerow No. 1 crowned by bushes and No. 2 lined with trees. Final score was trees 3, Airplane 1. (Ed's Note: It was rumored that when Lt. Ray reported to

The Story of VP-63

headquarters in Dunkeswell, England, he walked into the Commandant's office and said, "Lt. Ray reporting for duty with my plane, Sir," and handed him the wheel off the yoke of his disabled plane.)

The French, having waved us goodbye at takeoff, now hastened to greet us anew. One thoughtful person brought a bottle of brandy and a jerry can for some gas, of which we still had 300 gallons or so. We set a guard and returned to the village. Lt. Ray somehow received instructions to strip the plane of useable and classified gear. For this, he selected his regular crew members. Two days later, a truck arrived from Brest. I was placed in charge of two officers and three men and told to get Dunkeswell "howsomever". "Howsomever" included waiting several days in Brest for three minesweepers to clear for Cherbourg, a trip across the Channel in a Canadian corvette to Portsmouth, and then via British rail to London, and finally Dunkeswell.

* * * * *

TRANSPORT COATS: As I remember, to be really something in VP-63, you had to be a PPC, but beyond that you had to have a transport coat. Transport coats were not only prestigious, but you could tuck a bottle of bourbon in each pocket and it would be completely invisible.

Lt. Johnny Elliott and I went up to Dunkeswell from London by train. While at Dunkeswell, we teamed up as PPCs and usually went on liberty together. It was winter and we always wore our transport coats. We frequently used the trains and were just as likely to ride in the engine as in the cars. To ride in the engine you had to befriend the driver. Usually this was done by offering him a pull or two on the bottle in the pocket of the transport coat.

One day we boarded a train at a town called Taunton bound for a place called Weston Supermare. For some reason I got in a compartment, but Johnny went up to befriend the driver. When the train pulled out, I

MAD Cats

became concerned about his absence; however, there was no cause for alarm. Johnny was driving the train, having assured the driver of his previous experience on the Lackawanna and Southern, or whatever railroad runs through Kentucky. The bottle may have had some influence on the driver's lack of concern and his brief cockpit checkout on and off throttle, etc. Anyway, Johnny's green fore and aft cap and a rear view of a transport coat could be seen leaning from the engine cab.

The train we were on was a "local" and was intended to make frequent stops, preferably at stations with platforms, ticket booths, etc, Johnny's checkout did not apparently, include deceleration factors, braking distances, or anything else pertinent to stopping at stations. We soon approached the first stop, having gained considerable momentum and, as is customary, the brakes were applied; however, we went right through the station – in fact, we went right through the whole town and came to a stop in open country. Rumor has it that the driver took the whole thing with rare good grace, if not total hilarity, but Johnny's career with the British railroads came to an abrupt end.

* * * * *

PIGEONS: In a manner of speaking, you might say that the pilots of VP-63 could be generally divided into two groups. The first group included all those who experienced all things good and bad at Pembroke Dock. The second group included those who didn't belong to the first group.

The first group will recall a number of things including: powdered eggs, warm beer, cold quarters, London, hairy night landings with no "overs", cabbage, Brussels sprouts, and PIGEONS!

Did we eat pigeons? *Au contraire*! They were practically our last resort in the world of WWII communications. Certainly all members of the first group remember the pigeons.

The Story of VP-63

Included in the flight gear of every combat was a container designed to carry two homing pigeons. If you were forced down, you were to make certain that the pigeons went aboard the life raft so that, as fitting the occasion, you could write a message on the tissue paper provided, slip it into the little cylinder on Birdie's leg, and turn him loose. We gained confidence in the ability of those birds to fly home.

The was an old RAF sergeant at Pembroke Dock who was in charge of the pigeons. Like other flyers, pigeons had to get in so many hours in order to keep current, so to speak, on their inherent homing tendencies. For that reason, Old Sarge would ask from time to time that we scribble a dummy message somewhere out there on patrol, load up old No. 23, or whoever, and stick him out of the waist hatch (head first), and let him go. If he or she flew dutifully home, I presume the reward was an extra ration of bird seed. (Ed's Note: Or alky and grapefruit juice, eh Fairlie?)

The following story within a story cannot be considered to be factual or even legendary; but might be classified as folklore. It seems that this anonymous crew was out there in the Bay of Biscay one day in particularly nasty weather. About eight hours into the flight, they were, it must be admitted, totally lost. Radar down, no celestial, bad compasses, no bearing, etc. What to do? Someone came up with an idea. The pigeons! the coop was brought up to the cockpit. If these knew the way home, perhaps they might point their heads in the right direction. All that would be needed was to come to a course such that the pigeons were looking straight ahead and the rest would be easy! Alas, perhaps because of their confinement, there was confusion. The pigeons refused to point in the same direction. If there had been three pigeons, might there have been a majority opinion worth the risk, but with two divergent indications, one might fly fruitlessly toward the Azores, Newfoundland, or whatever. At this point, visibility improved and a new tactic was conceived. Let the birds loose just as if they were on a proficiency flight and simply follow them home!

MAD Cats

Most people prefer stories with happy endings. This one, like most operas, doesn't end happily. You might say as an analogy, the fat lady didn't even get to sing.

From here on, we can only surmise what happened. Everyone in Group One would concede that the PBY is, at best, a slow aircraft; with all that extra stuff hanging off the wings, you could take another ten knots from slow. Our Sarge reported the next day that both pigeons were safely home. Therefore, we must presume that they literally out-flew the aircraft and disappeared, leaving the aircraft to get home on its own. The aircraft was never seen again.

* * * * *

TRINIDAD: It was either late '42 or early '43 and we were all speculating on where the squadron would deploy. On this particular day, Captain Wagner summoned all of the officers to the ready room and had an armed guard posted outside. Having locked us in, he started the conversation with something like, "What I'm about to say is not to be repeated outside this room".

When we secured for the day, I drove to my apartment at Berkeley. My wife heard my steps on the stairs to the third floor, opened our door, and said, "I hear you're going to Trinidad!"

* * * * *

Lt. John Fox retired from the Navy at Great Lakes, Illinois, in June 1973. He immediately moved to Colorado Springs where he now resides and works in the real estate business. He says he is married to the same wife, has no kids, has most of his teeth, and has some traces of hair on his head. Over the years, he says he has managed to run into Mike Vopatek, Jeff Baker, Shot Lingle, Curtis Hutchings, Eddie Wagner (for whom he worked in the Pentagon), and Charlie

The Story of VP-63

Merriman. Charlie and he were together in Naples and afterward gave him a lot of support when he to Sigonella, Sicily.

****

Gibraltar in 1944. Source VP-63

Lt. (jg) J.C. Fox

ANOTHER SUBMARINE TO OUR CREDIT?: I joined VP-63 in Port Lyautey when I ferried in one of the PBY-5A aircraft. I had been in North Africa in VP-92 and we had 5As. The thing I would like to pass along to you hardly falls in the category of amusing, but it is interesting. It involved another plane commander, Elmer Mooney. I have never seen this written up in any WWII stories. It is a story that I hope you can check out with others in the Squadron, especially Mooney.

Elmer Mooney and I were out on a routine patrol on the barrier in the Strait of Gibraltar in February 1945. It was quite windy that day from the west/southwest. I got a 'hit' on the south turn and started working a clover-leaf pattern, dropping lights on each pass, and had worked up a pretty good line. Due to the high wind, I wasn't sure how much

MAD Cats

relative motion was due to the wind and currents, and how much might be due to sub motion. The position was, if anything, just inside the three-mile limit just off Tangier. No sunken ships were shown to be there on our charts. To try and resolve the question, I called in ol' Mooney to come down and work the contact for awhile and I went back to the barrier. He got the same results I had, so just before leaving the barrier, I went on over and made a drop of retro-bombs. If I remember rightly, one or two of the bombs went off. I can't remember if Elmer dropped the bombs or not.

We debriefed with the Fleet Air Wing A-1 types and never thought much more about it.

Some time later, before the Squadron left Port Lyautey, one of the A-1 people told us that they had a report from British Intelligence to the effect that a U-boat had surfaced in front of a convoy headed to the Strait from the northeast. Needless to say, shots were fired and the sub was sunk with many survivors. When debriefed, the sub Skipper has said that they were lying on the bottom off Tangier waiting to bag the *USS Augusta* (?) while it was on its way from the Yalta Conference carrying Franklin Roosevelt. The Skipper of the sub had pleaded with the British to tell him how they knew where he was and how he was detected because they were waiting in a no-noise mode. He said he had suffered too much damage to carry out his mission and had to finally surface. It was only coincidental that he appeared in front of the convoy.

I don't know if the Squadron ever got credit for it or not. I never heard anymore about it. You might want to check it out with Elmer Mooney and see what he recalls.

(Ed's Note: In Alfred Price's article *Aircraft vs. Submarine* which M.B. Cummins so kindly afforded me excerpts of regarding VP-63 operations, on page 196 appears the following quote: "During the remainder of March '44, three U-boats passed successfully through the Strait of Gibraltar, in each case by <u>sticking to the shallower water on the southern side</u>."

The Story of VP-63

It seems feasible that due to earlier successes of the subs to pass in the shallower water on the Tangier side, that it must have been considered a safe haven for a sub going after the President of the United States. VERY INTERESTING!

FOLLOW-UP LETTER FROM LT. MOONE: According to my flight log, I was on patrol on the barrier on the 8th, 10th, and 16th of February, 1945. I didn't find any notations about making a drop of bombs on those days or any other, for that matter. I distinctly remember a drop on the flight described.

FOLLOW-UP INVESTIGATION OF EVENTS CONCERNING YALTA CONFERENCE: The Yalta Conference lasted from February 4 - 11, 1945. Roosevelt left the Conference on the evening of February 11 and motored to Sevastopol, Krym for the night. Although Roosevelt's itinerary does not appear in the history books, it seems that he flew from Sevastopol to Malta, then boarded the *Catoctin* for the night. He visited Egypt for a day, then flew back to Malta and boarded the *USS Quincy* for the return home.

Since Malta is about 1,050 miles from the Strait of Gibraltar, the *Quincy* would have been in the vicinity of the Strait about February 17, the day after Lt. Mooney dropped his bombs, which could have been on the patrol of February 16. If the exact itinerary of Roosevelt can be located, a substantiation or repudiation of a "likely hit" by Lt. Mooney, and another sub victim from the MAD Cat's retro-bombs, could be made.

It is not known how long the sub was under water after it was damaged, or whether the sub surfaced in front of the convoy in the Atlantic or the Mediterranean. At any rate, the sub could have been hit by Lt. Mooney's bombs on any one of the days he was on patrol in the Strait - the 8th, 10th, or 16th. (Editor)

MAD Cats

Lt. C. A. Benscoter

<u>LONG LOST BROTHER FOUND</u>: My fondest recollection of WWII happened a few days after the conflict came to an end in the ETO (European Theater of Operations).

My brother received his flight training in Canada, after trying unsuccessfully to get into the Navy and Army Air Corps. After the US got into the fight, he transferred to the Air Force and got assigned to a squadron that went to Africa after the first landings there. He was flying B-26s when he was shot down. I guess he never did know what hit him – flak or a German fighter. They put the airplane down in a dried up lake, and successfully avoided being captured for a few days, but eventually the Arabs sold them out and the crew was captured. A few days later they were in a stockade in Germany, where they remained for the balance of the war.

You may remember that Cmdr. Brown had assigned me to the detachment in England where I had been since January 1945.

After Gen. Patton's Army had released most of the POWs during his run across Germany, I paid close attention to the BBC news, all the papers I could find, and any other information concerning released POWs. Eventually there were reports of the ex-POWs being collected along the coast of Europe to be shipped back to the U.S.

One day I obtained Admiral McFall's approval to take a shot at finding my brother. So I got my crew and the PBY-5A cranked up and drove over to Le Havre. We bummed the use of a vehicle and went to the ex-POW camp known as *Lucky Strike*. (There were seven of these camps along the Belgium and French coasts collecting ex-POWs for return to the States; all named for cigarettes.) There was a Major in charge of this camp of 75,000 ex-POWs. When I told him I was looking for my brother, he must have had a good laugh, but told us if he were in *Lucky Strike*, I would find his name on one of the shipping orders in the next tent. We were welcome to go and look.

The Story of VP-63

In the tent were stacks of paper all over the floor. They were about 18" high. The crew was still with me, so they all turned to. Within about ten minutes, one of the crew members located a Dan Benscoter on the shipping orders.

So back to the Major. Where could we find him? He took a look at the orders and said, "You go up this street four blocks, turn left for two blocks, and you will probably find him in the chow line." We did as we were told and, sure enough, Dan was in the chow line! Not only was he there and in relatively good health, but the former manager of my high school basketball team, Evald Benson, and another good friend of mine from the same metropolis in northwestern Pennsylvania, Bob Butler, were also there.

Next question for the Major; how do we get these characters out of here? He said he would redline Dan and Evald off the shipping orders, but we would have to go down the coast to get the Colonel's approval to take them back to England. So, we loaded up the bus with two happy ex-POWs and the crew back to the airport and the old PBY takes us to ex-POW Camp Chesterfield. The Colonel had no problem with our plans and gave the two ex-POWs ten days to report into a casualty detachment in London. Evald had been shot down while flying in the Eighth Air Force, so was happy to have the opportunity to get back to London; Dan had never been there, so it was a good trip for him.

Evald played co-pilot for me on the trip back to Dunkeswell, and that redheaded Swede had a grin from ear to ear all the way. We set them down in the Navy mess for a steak and a cereal bowl of ice cream, but they couldn't cope. The time in prison camp had had an effect on their digestive systems, so they just could not handle a large meal.

We played some golf along the south England coast the next few days, but Evald and Dan were eager to get into town for whatever they had been missing, so they took the train into London. The pictures of some of the golfing include Clint Rhen, a member of my crew at the time.

MAD Cats

THE SEQUEL: As you remember, the England detachment rejoined the squadron in Port Lyautey shortly after the war was concluded in Europe, and then it was not long until we were all headed back stateside via the south Atlantic. We all finally arrived in Norfolk amidst speculation of a Pacific assignment for the squadron. The squadron, however, was decommissioned. I drew an assignment in Pensacola, but with some intervening leave. Dan, in the meantime, reported in to the appropriate office in London, took a slow boat to New York, and a train home. Can you believe we arrived in Kane, PA (population 5,000 and a few) on the same day - July 4, 1945!

* * * * *

A PBY CAN TAKE ANYTHING, INCLUDING FLAK: On the same trip the Lt. Ray had his disaster in France, Lt. Fred Lake's crew flew over the Channel Islands, then German held, and took several flak hits. They were in the clouds at the time. One hit about at the intersection of the vertical and horizontal stabilizers. They were lucky to get to Dunkeswell after that one. Fortunately, the personnel on board had a lot of clothes in duffel bags stashed throughout the hull. The clothes took the shrapnel and, to my recollection, no one was hurt. There were lots of holes to be patched.

Incidentally, the British sent me an Order of the British Empire (OBE) about a year or more after I was released from the Navy for my attendance on their fair isle. It must have been for attendance, for we didn't accomplish a helluva lot. The trips into London were great, especially the theater.

Ens. L.R. Richardson

TROUSERS AND BILLFOLD: At the end of WWII, I was assigned to a crew to fly one of our PBYs back to the States. Our first leg took us to Bathhurst, SW Africa. We stayed overnight and prepared for

The Story of VP-63

takeoff at daylight the next morning to fly to Natal, Brazil. Lt. Elmer Mooney was the PPC and Lt. (jg) O'Brien was the co-pilot. I was the navigator.

When I entered the aircraft, I changed into a flight suit, hanging my uniform with billfold, etc. in the bunk compartment. The water in the Gambia River was very rough that morning, and with the load we were carrying, it was obvious that a takeoff was going to be difficult, to say the least. As we almost got to flying speed, the aircraft began to porpoise. We've all gone through this experience and know this may be said to be one of the most difficult situations one could encounter in a PBY. Anyway, each time the aircraft bounced back on the water it was more severe. On about the fifth time I noticed water coming through the port blister. On the sixth contact with the water, the aircraft bottom collapsed in the bunk compartment, and for a moment you couldn't see the after-station.

Not being able to get an answer from the cockpit on the intercom, I pulled my way forward and cut the throttles, which were wide open, the pilots being unaware of the havoc to the rear. I then headed to the after-station and by the time I reached it, the water was within a foot of the top of the hatch. Both rafts had been launched and I ordered flares fired for the rescue boats. I then asked for a head count. On being told that everyone was out, I thought of my billfold with $100 cash and a small money order that I had in my uniform. (You were allowed to bring $100 in cash out of the country and the rest had to be in a money order.)

The plane had stabilized with about 10" of airspace below the hatch door. I entered the aircraft and, with my feet, located the trousers and worked them gradually above the surface of the water. As I was leaving, I heard a low moan and saw Ens. Dean lying on the bunk in a semi-conscious condition with only his nose sticking out above the water. I dropped my trousers and called for help to get him out. He was very stiff and we had difficulty getting him out of the plane. The last bounce had crushed the bunk compartment area and Ens. Dean had taken the brunt of the impact. We finally got him into the life raft

MAD Cats

and subsequently the rescue boats picked us up. (Ens. Dean recovered from this experience in good shape, thanks to the luck that I had left my billfold in the plane.) The British attempted to tow the aircraft to the beach, but it took on water and sank.

The following day, the current had moved the plane so that only the tail was visible. All enlisted records were locked in the confidential locker. Divers were requested and dispatched from North Africa to recover them.

A life ring was anchored near the aircraft in case it moved with the current and submerged from view. When the divers went down, they found the plane sitting on a ledge which made it too dangerous to enter. Since the plane was in the seaplane landing area and was a hazard to landing aircraft, the British ordered it destroyed. Divers set the charges and detonated them from a boat. After the debris settled, they noticed something on the life ring. It was my trousers with my billfold visible and intact. As I had already returned to North Africa, it was necessary to send the billfold there, which they did.

The ink on the money order was not visible, so I requested that it be reissued. By the time I received the reissued money order, I was stationed in Columbus, Ohio, ferrying new Curtiss Helldivers. The money order was made out to Norfolk, my original destination, and could be cashed there only. I arranged a flight to Norfolk and my money order was in my flight jacket. It was stolen the night before I was to depart to get it cashed! I guess it just wasn't destined for me to cash that money order!

Lt. R. C. Spears

ENGINEERING GOLFER: Lt. Spears has been spending some time in St. Andrews, Scotland and England with his son and brother. He has been playing golf two or three times a week.

He retired from the Navy as Commander in 1962 and went to work for Aerojet General. From there he went to Washington as Staff Director

The Story of VP-63

of Professional Engineers, then to Baton Rouge where he has been Executive Director of the Louisiana Engineering Society for ten years. He has retired now, but keeps an office which he visits once a week as a consultant.

"Doc" says that Sniffer passed away in 1950. His aunt and uncle passed on and he has but one brother living in Arkansas. "My aunt was *Aunt Minnie*", he reminds us, "the name of the best crew in VP-63." That was the plane that Lt. Tanner was shot down in. (Ed's Note: And all these years I thought "Doc" was going back to Arkansas, do some farming and coaching, and every chance he got to eat nothing but okra and strawberries.)

* * * * *

NEED PETROL? After landing at Reykjavik, a British refueling crew yelled to a plane captain, "You blokes need some petrol?" The answer coming back was, "No, but we damn well need to gas up!"

* * * * *

SLOW DOWN: A personnel officer on a ship taking the ground crew of the squadron at Pembroke Dock. Noticing the greenish hue on the faces of the landlubbers and going to the Captain on the bridge while the ship was making full speed in rough weather and in sub wolf-pack area, "Captain, can you slow down? My men are getting seasick."

KENNEDY? Returning from convoy escort, a PBY making impromptu landing in the bay in front of Joe Kennedy's Hyannisport place, the plane having been blown off course by high crosswinds, the radioman opening the hatch and asking a soldier and a lady on the dock, "Where the hell are we?" The pilots, embarrassed by yelling, telling him to shut up! (The soldier was quite likely the elder son who lost his life in combat.)

MAD Cats

WILLIAM THE CONQUEROR: One crewman observing a Pembroke Dock beaching crew laboriously hauling a *Sunderland* up the slimy ramp with 20 men pulling on a line yelled, "Why in the hell don't you guys use that tractor off the side there? You haven't had a good idea since William the Conqueror!" (Ed's Note: Slight exaggeration!)

* * * * *

COAL PILE: The midnight raids on the coal pile at Pembroke, the coal was not to be used until December 1st, and the British finally posting a sentry to prevent thievery by the overfed, over decorated, and, worst of all, "over here" Yanks. The Maintenance Chief hoisting an American flag over the Pembroke barracks to stop British MP's from checking our quarters for stolen coal, and standing there in the doorway with upraised arm giving them the word, "See that American Flag? This is American soil!"

EASTER EGGS: The errant and much delayed package of Easter eggs sent to a plane crew member via Murmansk opened inside a Quonset hut. EXIT ALL HANDS!

J. A. Fahrner, Gunner

THE BLACK SHOE NAVY YEARS - PEARL HARBOR TIME: As a member of a USNR Surface Unit, I was ordered to active duty on May 18, 1941. After the usual confusion, such as sending a training unit to Boot Camp, we were assigned to man the *USS Republic*. This ship had been recently acquired by the Navy and was to be used in transporting troops. Built in 1904 by the Scots for the Germans, she was taken over by the Americans early in WWI. Her decks were reinforced so 4" or 5" guns could be mounted; smart long range planning by the (Germans). The *Big R* was sent to Iceland in the early fall of 1941, with the first American troops to be stationed in that

The Story of VP-63

country. Upon returning to the States, she was transferred to the West Coast for war fitting, with her ultimate destination the Naval Base at Cavite in the Philippine Isles. It was necessary for us to go to Pearl Harbor for the rest of our guns and munitions along with six bodies. The bodies were Philippine Commissary Stewards (USN) whose religion required them to be buried in their native soil. They were stowed in the hold along with the torpedoes. Of course, we had live bodies too, 3,500 infantrymen who were to strengthen the garrison in the Philippines.

The *Big R* weighed in at 29,000 tons and had to be anchored in Battleship Row where she lay until December 6, 1941 at which time she set sail on a southwesterly course for the Philippine Islands. On December 7, while our Captain read a notice stating that the Japanese had attacked Pearl Harbor, we could hear the ship's telephone ringing in the background. It was only minutes before we, the Ordinance gang (3 rates and 4 strikers), were made aware that our main line of defense, a 5" gun, was jammed in the train position. Not only was the gun jammed, its barrel was loaded with cosmolene – that heavy, hard grease preservative applied after WWI – and now it had to be removed and the gun repaired; how, I'll never know. The fan began to revolve very slowly as that stuff was poured into the blades.

Our Captain decided to utilize the passenger officers aboard to assist with the defense of the ship. They formed an organization called the American Ship Security Society, better known by the acronym ASSS. This group was headed by an Army chaplain. In their infinite wisdom, they decided to use whatever ordinance the Army had aboard to augment the ship's fire power. Twelve French 75s, field artillery guns from WWI, were taken from the hold and mounted on the lower deck with an ample supply of shells. These guns could only be depressed in elevation a few degrees and would have to be manhandled into a train position, just as John Paul Jones' men did 163 years before.

Then the ASSS decided we would mount what was laughingly known as anti-tank guns on the deck. These, too, were from WWI and were 37mm with zero degrees of depression and perhaps 10 degrees of

MAD Cats

elevation. A pile of shells was strewn about the deck in the general area of the guns. So far, we could generate one hell of a lot of noise, but couldn't do any damage. Come to think about it, this isn't true. If we were tied alongside another ship our size, we could have rendered utter havoc if it had been a wooden ship.

This charade is not yet finished. The ASSS decided that 200 infantrymen were to be assembled on the boat deck (second from the top) and upon signal from their officer, 100 men would climb the ladder to the sun deck. There they would fire five rounds from their .30-06s at any aircraft flying by. They would then run downstairs and be replaced by the second soldiers who would go through the same routine.

Many years after the war was terminated, by mutual agreement, a Congressional Committee, along with the Joint Chiefs of Staff, decided the procedure was an effective anti-aircraft defense. Not from the fire power, but any (Japanese) pilot flying over the *Big R* would have lost control of his aircraft and dove into the sea from laughter.

You wonder how this story concerns VP-63? Well, the author, when he was finally released from the "HOME", was assigned to VP-63; but that is another story and just may appear in VP-63's 1984 Journal. So, when you see a big fat CWO from East Buffalo ambling down the street with a destroyerman's gait, a funny twitch to his head, and a wild, but vacant, look in his eyes, please remember that happened after he became a part of VP-63.

Quonset Point, Rhode Island
Source: VP-63

The Story of VP-63

R.J. Watson, CY(AA)

<u>FROM LT. (JG) TANNER TO PEARL HARBOR</u>: VP-63 was one of my favorite outstanding experiences during my Naval career. It really started with the trip by Army transport ship from Pearl Harbor to Alameda with about 300 Navy personnel aboard, including Lt. (jg) Bill Tanner (only Naval officer aboard) and me. We worked together endorsing all orders so men would not have to go to the receiving station. Then on September 19, 1942, Lt. (jg) Tanner was VP-63 Personnel Officer, my boss. After the Squadron was commissioned. I found I was the only Yoeman aboard. The Squadron was allowed to have four. My days were long and hard; someone must have noticed, because I was promoted to First Class on October 1, 1942.

I remember at Quonset Point the uniform change from blues to whites. The first day I walked into the office with a Second Class rating badge on my uniform, the Skipper saw it and said, "I thought I made you First Class several months ago." I said, "Yes, but I forgot to change white uniforms." He asked when I was eligible for Chief and I told him whenever he wanted to send a recommendation to BUPERS. He told me not to change rating badges, but to get a letter off to BUPERS which was mailed at Iceland. When we arrived in Wales, the authorization was there; on August 1, 1943, I became Chief Yoeman. Then a trip to London to get uniforms. The Naval Attache had to give me the necessary authorization for clothing, shoes, etc. The uniforms had to be tailor made. For a cap, they used an RAF cap and made covers for it. The beak on my cap was much longer than that of any of the other chiefs.

I had tonsillitis at Alameda, Elizabeth City, Quonset Point, and Iceland. I finally went to the U.S. Naval Hospital in Roseneath, Scotland, to have the rascals removed. After weeks of RAF food, I arrived at the hospital at 1400 hours, after lunch, and had the cooks fix me ham and eggs and warm blackberry pie. When I had ice cream and milk for the next five days.

MAD Cats

Chief Watson had quite a Naval career. While at Notre Dame in a Navy VP-12 program, his roommate was Jackie Cooper. When he flew to Hawaii with his wife and two sons in January 1950, his pilot was Lt. Cmdr. William F. McSharry. In addition to his tour of duties at Pearl Harbor and with VP-63, he saw duty in Roan, Marseille, Palermo, Detroit, San Diego, Hawaii again, Salt Lake City, Kwajalein, again back to Alameda, Treasure Island, on the *USS Blue*-DD744, and the *USS Mansfield*-DD728. He retired at the U.S. Naval Receiving Station in San Francisco in June 1959.

Among his highlights with VP-63 were the following:

1) Working with Irby to produce our first anniversary commemorative booklet. Irby brought a poem to my office and wanted me to make copies. Instead, we worked together to come up with the booklet. The great man, without seeing anything, for the equivalent of $5, paid for 3000 copies. I had the pleasure of going to London to arrange for publication. Later we had to order 150 copies.

2) Our Commanding Officer at Change of Command ceremonies in Pembroke Dock with tears streaming down his face, shaking hands and saying goodbye to every officer and enlisted man.

3) Our new Commanding Officer doing everything expected of a Commanding Officer of our great squadron.

4) On Sunday, December 7, 1941, I was sitting on my bunk on the screened lanai (porch) on the second deck of the barracks on Ford Island. Several of us in our skivvies were shooting the bull when we heard aircraft approaching from the opposite side of the building. Someone remarked that a carrier squadron must be coming in. When the planes came in view, we still were not aware of what was going on until we saw bombs floating through the air and then three of our four hangars exploded in flames.

I dressed faster than ever in my life and headed for the mess hall on the first deck. We were told to get under the tables.

The Story of VP-63

Then groups were given rifles and sent to the roofs of buildings to shoot at the planes. The rest of us were sent to another building which was under construction - four walls, but no roof.

When bullets started ricocheting around, I made a dash for the Administration Building. Once there, I stayed at the side of Rear Adm. Patrick N. L. Bellinger - I was his secretary. We would go from one side of the building to the other side, helplessly watching the devastation. The Admiral would remark, "Look at that son-of-a-bitch make that beautiful torpedo run" and "That bastard made a perfect bombing run."

It so happened that on Monday, December 1, 1941, Adm. Bellinger, two of his staff officers (Cmdr. Ramsey and Lt. Cmdr. Coe), and I boarded the *USS Curtiss* and got underway for Hilo, Hawaii. For three days we had training exercises with the PBY seaplanes from our patrol squadrons practicing torpedo and bombing runs at the *Curtiss*. On Wednesday, we flew back to Ford Island in the PBYs. On Sunday, six days later, we were watching the execution of what we had been practicing.

The Naval Air Station Administration Building was not hit, but one bomb, a dud, landed by the flagpole in front of the building. The *USS Curtiss* has returned to Pearl Harbor Saturday afternoon, but was hit during the attack on the seventh.

F. C. White, ARM

SWITCH OVER, REID! I remember being half-scared to death the night we flew PBYs from San Diego to Kaneohe, Hawaii. I was in Lt. Reid's crew (now Captain Reid, retired) and we were about midway on the flight. One of our engines cut out for a few seconds, but due to the alertness of our mechanic in the tower, Eric Engman (I believe), he kept the engine going with the wobble pump. I think what happened was the tank went dry before it could be switched to the other one.

* * * * *

UPSET, BIG BEN? After I was in Lt. Reid's crew, I was in Cmdr. Hutchings crew in England and Port Lyautey. I went into Lt. Cmdr. Benscoter's crew when a detachment of us went back to Dunkeswell, England. I was up there when the war in Europe ended and some of us got looped on de-icer fluid and grapefruit juice to celebrate the war's end. Lt. Cmdr. Benscoter got a little upset with us. But, then, a MAD Cat would do almost anything.

PBY Cockpit. Source U.S Navy

The Story of VP-63

J.W. Carthel, S2c

FRIEND OR FOE: I sailed for North Africa from Norfolk on Sunday, March 3, 1944. We shipped out in pea coats and tropical pith helmets (to confuse the enemy as to our ultimate destination). That Sunday in early March was a very frothy day and, of course, our friendly company cooks fed us greasy porkchops. As expected, I joined the others at the rail.

The next day, not so fully recovered, I was riding a seat on one of the trough-like salt-water heads. An explosion suddenly occurred and the salt water splashed all around me! I headed topside, buttoning en route, to escape the torpedo or whatever. The ship's company of the *USS Albermarle* - Sea Plane Tender (Able Mable) had failed to tell the passengers that they were going to practice with the 3" deck gun. Was that crew on that old sea plane tender really on our side in March 1944?

L.P. Harris, ACRM

LOW FLYING: Soon after our arrival in "Blighty", we were sent on a familiarization flight around the south of England and were caught out in one of those British fog banks. In an attempt to reestablish contact with the ground, CAP Smithee was in the bow in helmet and goggles peering over the side as we descended through the soup. When he began waving his arms and screaming, we knew that we were in close proximity to the ground. In fact, it looked like we were flying through some farmer's apple orchard. I remember, also, Hickey and Beam being in the crew. (Ed's Note: So that's where you got all those apples, Smithee.)

Then there was the morning we were making a pre-dawn takeoff toward Milford Haven where a convoy had taken refuge during the night. They were anchored across the mouth of the bay with barrage balloons displayed, a virtual fence in front of us. We had to leave the estuary and go clawing for altitude across the fields. The farmer's daughter later asked if we would be so kind as to not fly so close to the house, as her father was afraid we would bring the chimney down. I tried to explain that we were concerned about the same thing and she could depend upon our cooperation.

* * * * *

<u>PEARL HARBOR DAY</u>: On this day I was a member of R-12 stationed across from the island of Kaneohe Bay. We weathered two attacks. During the second attack, I took refuge in the first hangar while the (Japanese) were making their bombing run. I was inside the hangar in the stairwell when the salvo came through the roof. Soon after that, while the parked planes were burning and their ammunition exploding, someone running through reported that a chief, AOC Grisham, was badly wounded out on the hangar deck.

To my eternal amazement, I felt compelled to rush out and drag him into the stairwell where we applied a tourniquet, made from my undershirt, to his right leg, which had been badly mangled. When the raid was over, we loaded him onto a passing truck and sent him to the dispensary. Never saw him again, but in 1961, while I was stationed in Patuxent River, Maryland, I met the fellow who had helped me load Grisham on the truck. I also learned that Grisham had survived our treatment and was employed as a guard at the White Sands Missile Station from my division. The two had been life-long friends. Small world, eh?

* * * * *

<u>MEMORIES OF ENGLAND</u>: I still tell stories about being on reduced rations at Pembroke Dock, the early morning briefings, and the cage of homing birds we used to take along on patrol. Anyone remember them?

* * * * *

The Story of VP-63

M.B. Cummins, AMM-1/c

<u>THE SPERRY MAN</u>: Melvin Boyd Cummins, AMM-1/c, is a plank owner in VP-63 and said he had the honor and pleasure of serving aboard for all of her time of commission. You will recall Melvin as the "Sperry Man" who, along with others, was an instrument and auto pilot man. He was Plane Captain of the aircraft PPC'd by Lt. Woolley that made the initial contact on the German U-boat U-761, subsequently sunk with the assistance of Lt. Baker's aircraft and surface vessels.

Melvin retired from the Navy after 30 years as a Lt. Commander. His last duty was on the staff of Commander Fleet Air Alameda as Aircraft Maintenance Management Officer.

* * * * *

42nd <u>REUNION SUGGESTION</u>: Melvin thinks it's great that we are having a reunion of the Squadron after all these years – forty-one after the original commissioning of the Squadron to be exact. He would like to see a book put out next year with reunion pictures and information about what everyone has done since the war and where they now are located.

Melvin suggests a reunion next year be held at Alameda where it all began forty-one years ago. He says we could muster at the old hangars, tour the Air Station, go aboard an aircraft carrier and other ships in port, and cruise the Bay, especially under the San Francisco/Oakland Bay bridges where we used to fly our PBYs, etc. Sounds great, Melvin!

* * * * *

MAD Cats

F.L. Rosenmund, ACMM

<u>LT. PARKER'S ONLY FRIENDS</u>: Fred Rosenmund, ACMM, recalls the story about Lt. Sam Parker's problem in tying up to a buoy one fine day while stationed at Pembroke Dock. It seems that Sam made four or five attempts at tying up to the buoy amid swirling currents and tight quarters with other planes tied to buoys nearby. He finally made it, but not without damaging the wing tip float of an adjoining plane. When secured to the buoy, Lt. Hillie called to Parker and said, "The dinghy is coming, Sam. Are you ready to go?" He replied, "No, Hillie, I'll stay out here with these caged pigeons. They'll be the only friends I'll have."

* * * * *

<u>CANNED BOOZE</u> ? Then, he recalls the cans of pineapple juice, tomato juice, etc. that arrived in the mail one fine day at Port Lyautey. It seems some of the ladies in the States had used their ration stamps to purchase good booze and had it canned by some cohorts

* * * * *

<u>ICELAND INSPECTION</u>: Rosenmund recalls the first personnel inspection in Iceland as ordered by the Commandant there. He recalls that they had hidden their dress blues under the Quonset hut floorboards and pleaded to be excused on the grounds of not having the proper uniforms. They were found out and sentenced to punishment of two hours of military drill for three days under the watchful eye of a Marine sergeant to shape up. (Ed's Note: I'll never forget this inspection, but all the time I thought the reason for the motleyest looking group of men I ever saw was that all the uniforms were on their way to England. You learn something every time someone opens up their mouth, don't you?)

* * * * *

The Story of VP-63

Adam J. Chaisson, ARM

The story of the search for survivors of the *Coronado* is told elsewhere. It does not mention that in addition to our normal operating gear, we also carried our beaching gear and other gear as well (more weight) because we had been packing for our trip to Quonset Point. We didn't take time to unload the gear.

I was not on the flights from Quonset to Iceland. I got there via seaplane tender. I missed out on that harrowing experience.

Adam and Anna at the MAD Cats Reunion

FROM BOSTON TO ICELAND – SLOW IT DOWN: Having been bumped from the flight to Iceland by a higher rated radioman, I was forced to go by ship from Boston to Reykjavik. This particular ship had been converted to a sea-plane tender. Shortly after leaving Boston Harbor, we were told we would be travelling full-speed through submarine-infested waters, and if anyone should fall overboard, they were on their own to get to shore, because we would not be stopping to pick up anyone or anything.

The seas got rough with wind and rain and the officer in charge of our unit told the ship CO that he would have to slow down because all his men were getting pretty sick. His answer was that he would slow

MAD Cats

down only when he got to his destination. After we arrived at Reykjavick, the ship yeoman whom I had gone to high school with, told me that the cargo they were carrying – besides the group of men from VP-63 – consisted of, among other things, aviation fuel, contact bombs, depth charges, torpedoes, ammunition, and supposedly for the first time in naval history, a shipment of nitro glycerine in its ready-mixed form. I would say the skipper was in a hurry to get rid of all his cargo – including the people from VP-63.

Chaisson was on the original *Aunt Minnie* crew. The day *Aunt Minnie* was shot down, he was scheduled to fly; but at wake-up time was told that he was taken off that flight for some unknown reason.

The original crew of Aunt Minnie
Sept 17th, 1983
Ware   Knight
Gaskin,   Martin   Spears   Chaisson

Source: VP-63

The Story of VP-63

Lt. S. H. Castleton

SOME MEMORIES OF A JUNIOR OFFICER OF VP-63: It was the morning of September 19, 1943 that Bob Warren, Gil Knight, and I, Bill Craddock, drove out of the Corpus Christi Naval Air Base as commissioned officers in the US Navy Air Corps. For the last nine months of training as aviation cadets, we had been sworn at and kicked around by all hands, particularly by Marine sergeants and flying instructors. The reality of being Naval officers had not fully sunk into our skulls. This lack of reality would last only a few more brief moments.

As we entered the gate, several Marine guards jumped to attention and briskly saluted. Caught off guard, it took us a few seconds to respond with a salute. As we departed the gate, complete silence swept throughout the car. About a block from the entrance gate, we looked at each other and spontaneously broke into laughter. The irony of it, others saluting us for a change! We realized then that we had become Naval Aviators in one of the great fighting forces of the world. We were beaming from ear to ear with pride and satisfaction and determination that we alone would win the war. If we had known then what we were to later learn, we wouldn't have been so cocky.

This day, September 19, 1943, was the day of commissioning of Naval Patrol Squadron VP-63 in Alameda, California, and this was our ultimate destination after spending a few days back home.

Alameda and San Francisco were beautiful parts of the country, especially after the plains of Texas, and it was with great joy and anticipation that we were to join our first operational squadron at this location. With considerable awe and a bunch of respect, we met the officers of VP-63 who were to be our superiors and associates for some time to come. Anyone above the rank of Ensign commanded our immediate attention and Lieutenant Commanders were something out of this world; born, not made, with halos and gifts of fortune from the gods.

I particularly remember the first meeting with the skipper. Lt. Cmdr. E.O. Wagner, the Executive Officer, Lt. Cmdr. C. H. Hutchings, and the senior officers who were to be PPCs of the aircraft we were to fly.

MAD Cats

They all had lots of experiences on either or both ships and various squadrons in many areas; some had already experienced the fury of the war in the Pacific. We also had some experienced chiefs and enlisted men in the Squadron, and all of us fledglings were blessed to have this experience aboard. With the officers and men that we had, I think a sense of security prevailed among us that everything was going to turn out alright.

All of us will remember 'Skipper' Wagner as a straightforward, clean-cut gentleman, sympathetic and sincere, and dedicated to the Navy and the men under his command – a real leader for whom everyone wished to do the best he could. It wasn't long before personnel in VP-63 became known as "Wagner's Men". The very existence of VP-63 as a MAD Cat anti-submarine squadron throughout the War was due to Wagner's influence, determination, and undying faith in this new approach to anti-submarine warfare. Wherever and whenever the Squadron had success with the "gear", Wagner was there in spirit, smiling and triumphant in the background. All our hats are off to this great man – a true pioneer in Naval Aviation history.

Lt. Cmdr. Hutchings, the Squadron's first Executive Officer, had a first love. It was celestial navigation! He had written several books on the subject while instructing at Pensacola, and he always insisted that we underlings become proficient in star shooting.

I can remember a few stories about ol' Hutch, as we called him, when we had to spend a lot of time on the base at night shooting stars with an octant. It sort of got under our skin when this kind of duty kept being resurrected and the saying got to be commonplace among us underlings – ol' Hutch is going to make navigators instead of aviators out of us yet. After all, we would have been much better entertained fighting the "Battle of the Lemington Hotel", where all the beautiful girls met all the handsome Naval officers, then standing out there overlooking the Bay shooting stars. If we had known then what we later were to learn – that ol' Hutch was a former boxer and wrestler – this would have been a <u>second</u> reason why we should have shown more respect for him than we did on these occasions. I have to say, looking back, that ol' Hutch's insistence that we become competent in all phases of navigation paid us good stead over the years to come.

The Story of VP-63

Flying over the oceans of the world, often for as long as 16 to 20 hours at a time, we were never lost, even though we didn't know where we were a lot of the time.

I remember well ol' Hutch's story to us about the wonders of celestial navigation. One story goes like this: Ol' Hutch would leave San Francisco, fly all the way to Tokyo, never see land, use only celestial navigation, and place a bomb right square in the middle of the Emperor's Palace. Man, that's some navigating!

The Executive Officer's job was a tough one since he was the one to see that the Skipper's" orders were carried out. Many an Executive Officer was disliked by all hands, to say the least. But to Cmdr. Hutching's credit and to show how he carried the burden, he was sincerely liked and respected for his ability, his fairness his directness,and above all, for his love and dedication to the Navy, the "Cause" for which we were all fighting, and for the men who served under him. When he assumed command of the squadron, he carried on in the magnificent tradition set by his predecessor, Cmdr. E.O. Wagner.

Volumes could be written about our PPCs, junior officers, chiefs, and enlisted men, but suffice it to say that they were the greatest bunch ever put together in one squadron. I have yet to hear any of them express any regret at having been in the squadron, and all of them I have heard from say it was their greatest and most pleasant and memorable experience in all of their Navy wanderings. 'Nough said!

There were some characters in the squadron with whom I was, from time to time, more closely associated. These associations bring back memories of some of the things that happened that impressed me as either being human interest or comical. Some of these stories follow.

I'll always remember Stan Castleton, a cool and suave gentleman. As everyone knew, Stan was quite a ladies man. I know because I was the victim on many occasions, when going out with him,of getting the least attractive girl of the pair. Also, he was a real connoisseur of fine whiskeys and could tell you the brand of many of them just by taste. I've seen him beat bartenders out of many a drink by betting them he could name the whiskey by taste. Of course, he would badger the

MAD Cats

bartender into making the bet by loudly proclaiming his astuteness in the art of differentiation of the spirits. He was always cool, calm, and reserved gentleman, whether partying or on duty. Boisterousness was not in his repertoire. When he would come in from a night on the town, he would wash his face with cold water over and over again until I thought he would drown himself. He never had a hangover and he said the cold water treatment kept him from having one. He was as calm and smooth a pilot as I ever flew with in the War. He came in from out in the Bay one night in horrible weather, on one engine, and could not get into the base at Pembroke. Short on fuel, he cut across the mainland in an attempt to reach our alternative base at Hamworthy on the southern coast of England. He made it safely after flying right through the middle of a barrage balloon defense the British had strung up over a village.

PBYs were used as dive bombers in the early part of the Japanese War. This reminds me of the story on Ens. Gates when he was practicing maneuvers against theoretical submarines with MAD equipment. Effectiveness of the use of the MAD was greatly enhanced by getting to the point of contact as quickly as possible. This was not an easy chore for the slow and lumbering PBY, especially if the sighting was made when the aircraft was at high altitude. (The story goes that once a PBY pilot radioed that he had a submarine in sight on the surface and was attempting, with full throttles, to catch it but was losing ground – the sub was outrunning him.)

Anyway, Ens. Gates had his theoretical sub sighting and went into a dive to arrive as quickly as possible at the spot. Personnel on the plane said that the airspeed and rate of descent indicators went around three times, wings buckled, and rivets popped like popcorn in a hot popper when he pulled out at about 50 feet above the water. The gravitational pull of "g's" was so great that people's eyeballs almost popped out of their sockets. Those who witnessed this incident as well as personnel on the plane thought it would be the last dive; naturally it scared the wits out of everybody. Ens. Gates was encouraged not to ever try that stunt again, but I don't think he needed much persuasion as he, too, was pretty shaken over the incident.

The Story of VP-63

Our operational squadron departed Alameda on March 15, 1943, after six months of outfitting and training at Alameda. We patrolled the entire length of the Atlantic and into the Caribbean Sea for enemy submarines. We had small detachments at one time or another at Quonset Point, Rhode Island; Elizabeth City, North Carolina; Jacksonville and Key West, Florida; and Bermuda to name most of them. We patrolled the East Coast until June 22, 1943, on which date we set out for Argentia, Newfoundland, as our first stop on the way to Iceland. Admiral King had said he didn't ever want to put PBYs in a war combat zone after the way they were chopped up in the Pacific theater. I said to myself, "We're sure heading close to the war zone, but Iceland is still a safe distance away, but still plenty close for these old slow birds. I wonder why we are going to Trinidad via Iceland. Sure seems like a long way around." I didn't dream of what was to be in store for us later on as we started winging our way on to Iceland.

I will never forget this trip to Iceland and neither will a lot of others. We left on a flight and encountered terrible weather on the way. With the entire Squadron in the air, and in the kind of weather, some didn't seem to know where others were and there must have been a few near misses from the sound of voices over the radio. You would hear a voice blurt out over the radio, "Pull up! You're right over me!" Or "Get down! You are hugging me underneath!" Or "Pull to port!" Or "Pull to starboard!" They had men in the after-stations flashing the Aldis lamp around in the sky to give warning of where they were in relation to the other planes. Lt. Knight said they were never able to find an altitude where the weather was good in spite of reports to the contrary from other planes at various altitudes.

There was a funny story about Lt. Nicholson's trip to Iceland. It seems that he missed his destination at Reykjavik and made his first landfall in extreme northern Iceland, about 200 miles to the north. His plane captain kept advising him that they were low on fuel, so Nick had no other choice but to start lightening up by throwing out everything on the plane that wasn't nailed down. I think about everything went overboard except his radio, MAD gear, and crew. Well, when the plane finally landed at Reykjavik, Nick discovered on landing that he had plenty of gas left and that there had been no need to throw anything out of the plane to lighten up. He asked the plane

MAD Cats

captain why he had reported to him that they were running low on gas. His reply was that he was saving it for the wife and kids. You can surmise what "gentle" Nick did upon getting this answer! You remember the man without a country story, don't you?

A humorous story to some of us, but I'm sure a distressful event for others (especially Exec. Officer Hutchings) occurred in Iceland a few days prior to our departure for England. I believe a small contingency of planes had already departed for the new base at Wales, and Skipper Wagner had gone ahead of everyone to make preparations for our arrival there. All of our gear and clothes were enroute by ship except for a few clothes we would wear in flight. The events that occurred in downtown Reykjavik two nights before our departure aggravated an already sensitive admiral of the Icelandic Forces and resulted in an inspection of the motleyest crew that the Navy or any other military force had ever seen.

The custom at dances in Iceland was for the man to sit on one side of the hall, the girls on the other, and never seated together. As Lt. Spears, myself, and a few others left such an occasion a few nights before we were to depart for our new base, we couldn't help but notice what seemed to be hundreds of girls hanging out of windows in a hotel next door waving for us to come up. Well, we had had enough to drink to think it would be fun, so we made like we were going to climb the stairs on our way to the upper levels. We had taken only a few steps when four or five Shore Patrol confronted us. Some eager beaver among them decided it would be a good idea to call the Admiral and report a disturbance downtown by some VP-63 personnel - and this was at 2:00am in the morning. Needless to say, this didn't make the Admiral very happy since this seemed to be a repeat performance for some other misdemeanor that occurred during our brief stay on the Island.

As I stood watching the developments, Lt. Spears and a few others vociferously defended their rights to be standing on the street in front of the hotel. I think they succeeded in getting themselves on report. The next day we learned that the Admiral had called for a complete squadron inspection.

The Story of VP-63

When we gathered and lined up for the inspection the next morning, you would have had to have seen it to believe it. We could have been taken for a bunch of derelicts who had visited the Salvation Army and robed themselves with their discarded clothes. No one had anything on that was really neat, pressed, or matching. There were all combinations of different colored portions of the various uniforms - blues with whites, greens with blues, blues and greens with khakis, shows of every color, and even some flight boots. The reason was, of course, that we were moving to Pembroke Dock the next day and all our uniforms were in transit by ship to the new base. Really, this was the motleyest looking bunch of men ever seen at one gathering in anybody's Army or Navy.

I don't think the Admiral was aware of what was about to confront him, and you should have seen the look on his face when he entered the hall. He must have thought he was in the wrong place. If he had been in any mood but just plain mad, he would have fallen on the floor in laughter, but he carried out his inspection in true Navy tradition, line by line, studying each individual as he passed. I couldn't help but wonder what was going through his mind as he carefully and meticulously studied each person from head to foot. This was not one of the best times for VP-63. I'll wager that when the Admiral went home and described this scene to his wife, they both fell on the floor and rolled with laughter. Only the explanation that our gear had departed for England, and especially the act that we were going too, saved us all from the brig.

Well, we finally arrived at Pembroke Dock, Wales. This was a new experience for us, and an honor, to be based with the Royal Air Force that historically had made such a name for itself. We learned while there to never sell the British short, because they knew how to fight a war. They did some things the hard way, but they made do with what they had. Spare parts for their *Sunderland* flying boats were in short supply and, honestly, they were keeping those monsters in the air with barbed wire and guts. They would simply cut holes in the sides of the fuselage to mount guns to defend themselves against constant attacks by swarms of German Ju-88 fighter planes.

MAD Cats

The British were men usually older than their American and Canadian, and Australian comrades, more sophisticated and reserved, and usually with handlebar mustaches. This posture became completely obscured when they became inebriated. I can remember these stately men turning into kids at the conclusion of a dance when they were fully loaded. They would place a large padded chair at the end of a long room and run as fast as they could, turn a flip just as they arrived at the chair, land in the chair, bottom and feet upward, turn the chair over, and come up standing on the floor on the other side. Lt. Jim Hardy recalls the party they had just before we arrived where the Wing Commander broke his arm. He must have missed the chair. Probably hadn't had enough to drink!

Nothing could be written about Pembroke Dock without including something about their mess hall. Although American steaks and the like were available at nearby Army facilities, Skipper Wagner concluded that it would not be right for us to be housed with the British, fight the same war, and not eat the food they did; and rightly so. This was the only way to go, although the decision would not have won an Academy award among the personnel of VP-63.

Well, you'd have to eat in their mess to fully appreciate how our accustomed standards of eating went downhill. Along with the food; which usually consisted of kidney pie, Brussels sprouts, hard knotted potatoes with no seasoning, and marmalade. I'll always remember the plates. It seemed that the British were very limited in soap and hot water in which to wash their plates and dinnerware, so they just lightly washed them off in cold water, which I don't think they changed very often. I know the towels they dried them with weren't cleaned very often and this didn't add to fragrance. The dishes had developed a rather rank odor and this didn't add anything to the food being served. As far as I know, no one lost much weight, but this was primarily due to tea time which was the only time most of us halfway enjoyed eating anything. But then, suddenly tea time ceased! Things really got bad then. We didn't find out what caused this until forty years later - at our 41st reunion!

I will always remember the situation concerning getting eggs in England, anytime! Only a few of us ever had an egg during our entire

The Story of VP-63

stay in Pembroke. If you saw someone slink into the mess hall holding something with both hands in his pocket, you immediately knew that he had a wonderful possession - an egg! Lt. Jim Hardy remembers the 5-gallon can of Catsup that Nick and Woody, proprietors of the *Busy Bee Cafe* located in his BOQ, mysteriously acquired. It lasted about 24 hours. Hardy says Spam isn't much without Catsup and other officers soon discovered their grill, complete with pans and coffee pot, and headed in that direction for evening meals. The one stipulation was that all guests supply their own grub. Traffic at times resembled the gala days of Times Square at New Years. Hardy recalls that we did get a steak when on patrol in the Bay and says he always had a slight (very slight) feeling of guilt thinking about what the guys were having for dinner back at the base.

P.F. Pfannstiel reports that the (British) were not great feeders. He said that the first meal he had in England consisted of two overgrown sardines, a couple of slices of dark bread, and some honey. His first breakfast in Gibraltar consisted of ripe fried tomatoes, powdered eggs, and something they called bacon. I'll never forget the odor in the enlisted mess when they were serving the fried kippers. Man! I'll never know how they stayed in there. They had to be awfully hungry. But this was war. England had been through a lot, and we were lucky we hadn't had the battles and hardships that they had already endured. The war was being won and that's what really counted. We really got enough nourishment, but we weren't about to admit it.

I remember well Lt. Rowland Fairlie who flew with Lt. Hardy and myself for a good part of our tenure with VP-63. One night it was reported that Fairlie bought over 100 Scotch drinks for the British officers; after that, the British thought he had "hung the moon". The British pay was so low that they could seldom afford Scotch - it was usually hot beer and Schnapps for them, the latter tasting like the laxative Feenamint, with both having the same results – for me and others, too.

There were a couple officers in our squadron, Lts. M.J. Simer and John Elliott, whom I'll have to tell a couple of stories about. There was never a more fun loving pair to hit the beach; and when they went on liberty together, they were full of vim and vigor and ready for any

MAD Cats

eventuality. They were both burly men, particularly Simer, but both had hearts of gold.

At Pembroke Dock, there were insufficient quarters on the base for all of the officers, so a lot of us lived up on the hill in two-story houses. Well, Simer and Elliott lived a block above us and had the reputation of having the best little bar in town. It developed that a certain British fellow got rather attached to the bar and would invite himself in for several drinks or more daily, even when the hosts were away on patrols. This kept us for quite awhile until Simer and Elliott decided they were going to have to take remedial action to deter his ambitious lust for the hideaway. So they got him all tanked up one fine day, put him on one of those British bicycles, and sent him off down the hill toward the base. People who saw him go by said they thought he was a streak of lightning. I think he went all the way, with His help, across the docks and into the estuary because someone reported seeing a bicycle walking on water. I saw him a few days later, limping slightly, but otherwise in good shape. I never did see him on the hill again.

Then there's the night Elliott and several others got into a drinking party with the base Provost Marshall – a Scotsman, highland kilts and all. Late in the party, Elliott convinced the Provost Marshall to change clothes with him for he was anxious to see what he looked like in kilts, long socks, and the Tartan. Though most of us did not get a chance to see this masquerade, we can all picture in our minds what he must have looked like. Well, some way or other, and no doubt preconceived, Elliott slipped off with the Marshall's outfit, leaving him undressed, so to speak, and not fit to be seen in public wearing a pair of pants and shirt belonging to some U.S. Naval officer. The Marshall had gotten so polluted that he didn't remember who got away with his clothes. The story goes that he stayed in his room for three days, mortified that he would be seen in public undressed without his kilts, tartan, and long socks. I understood that the Marshall finally got his right clothes back, but I've always wondered who had the nerve to take them back to him.

"Tea time" at Pembroke Dock was a sacred event occurring twice daily, morning and afternoon. If anyone came in from patrol at tea time or had some emergency, they were out of luck to get any

The Story of VP-63

attention. Many a day our flyers sat at the buoy waiting for this event to end. Nothing could be done during this period. It would be like interrupting the preacher in the middle of his sermon to tell him a shady joke. Sometimes it really got under the skins of those having to wait for this eventful occasion to end; like the time, as reported, that Ens. Smithee picked one up and threw him into the water. It wasn't all that bad, though, because those of us who were starving to death on British food became much better nourished at tea time. Then a terrible thing happened. Someone scuttled tea time and many of us went immediately into anemia. How could anyone do such a thing? It was like breaking all ten of the Commandments at one time. They should have found out who the culprit was, even if they had to hire Scotland Yard, given him a mock trial, and let him suffer the consequences.

Everyone remembers Lt. V.A.T. Lingle, III, the little redhaired, redfaced fella from Georgia. I guess "VAT" got more ribbing from fellow officers than any other officer, and it would take a book to relate all the humorous situations that evolved around him. Lt. Andy Reid, with whom "VAT" flew a good part of the time, had some good human interest conflicts involving him. I remember the time that John Elliott scared the wits out of him when he dressed up like an Arab sheik, sword and all, and rushed into his room swinging the sword wildly about and shouting something about "death to all traitors".

"VAT" loved his phonograph records, especially the ones about Mr. and Mrs. Pettibone, and he played them continuously, much to the chagrin of us all. We had heard the Pettibone record so many times we could sing them backwards. I'll always remember the part where Mrs. Pettibone comes into the room and stands in front of his wife "stark naked and twice as cocky". "VAT" would fall out of his seat with laughter every time he heard this, and we're sure he heard it a thousand times, or more. Lt. Andy Reid tells of the time when "VAT" was three minutes late in arrival at Alameda after a five-hour Nav flight. He said, "Ah! Thu navigatuh musta dropped thu octant!" He was referring, of course, to the press release by the Army Air Corps of how a South Pacific flight managed to get lost and dunk Eddie Rickenbacker, among others.

MAD Cats

I remember, as most of us do, Lt. Walter J. McHargue, a feisty, heavyset, likeable sort from Boise, Idaho – always eager and charging to every assignment. If he flew in the air the same way he behaved on the ground, he must have been a tiger. I would pity the poor Germans who ever encountered him.

One day we were playing baseball and Mac, as we called him, was playing the catcher position as he always did. When he got the ball on a pitch from the pitcher, he would likely throw it anywhere in his peppery way, except back to the pitcher. I knew this, but I was standing at the shortstop position fat, dumb and happy when Mac returned a pitch; not to the pitcher, but right at me, as hard as he could throw it. Of course, I wasn't looking, and it hit me square on the nose. If you've never played with a so-called softball, then I have to tell you the name is a misnomer; it's as hard as hell! I was embarrassed and mortified, as well as physically injured. If anything didn't need aggravating, it was my proboscis - it was big enough already! Well, my eyes watered and I saw stars for an hour. I'll never forget how everyone thought it was so funny, particularly Lt. Cmdr. Hutchings who almost fell on the ground in laughter. I never was able to understand why everyone thought it was so funny.

There was a top officer, I believe it was Lt. Cmdr. Platt, Base Executive Officer at Port Lyautey, who put "the fear of the gods" in everyone, including some of the brass. He used to come into Skipper Hutchings' office on occasions and, in doing so, had to pass by the Duty Officer's desk in a small adjoining room where several officers were always lounging around. One day he came through and no one gave him any recognition, including the Duty Officer. Skipper Hutchings noticed this and advised all Duty Officers to bring everyone to attention in the future when Cmdr. Platt entered the room.

Well, on this particular day, I believe Lt. Billy May was on duty at the desk. When Cmdr Platt rushed through, as he usually did, May jumped to his feet and shouted at the top of his voice, "ATTENTION!!!" His voice echoed throughout the building and all the hangars and the Arabs must have heard him in Casablanca, over a hundred miles away. It nearly scared Lt. Cmdr. Platt to death and he jumped about 4 feet in the air. He was already in Skipper Hutchings

The Story of VP-63

office when the call rang out, and I'm sure he also jumped out of his seat with furiously palpitating heartbeats. Officers lying around the anti-room fell off their chairs and one guy knocked over a small file cabinet in scrambling to his feet. Skipper Hutchings then advised future Duty Officers to tone down their "Attentions".

Sometime late in 1944 or early 1945, the Navy decided to send two dirigibles to Port Lyautey to scout for submarines. They were to sit at predetermined locations out over the ocean and watch for subs. When and if they sighted one, they were to call us in for the attack. Some Arabs had never seen these lighter-than-air balloons, and it was reported that many of them got on their knees and prayed to the heavens when the two airships crossed over Port Lyautey for the first time.

It so happened that the Skipper on one of those balloons was an admiral's son who, apparently, had little regard for values and did as he pleased most of the time. It was reported that rather than dump gasoline to lighten ships for landing, a routine custom, he would throw off radio gear, sensitive equipment, and anything else that wasn't nailed down. For some reason, he seemed to always get away with it.

They moored these monsters on top of the hill southwest of our barracks and you could see them floating and waving in the breeze when tied to the moors. One day a high wind bounced one up and down so hard it burst and lay spread out in a big heap on the ground. Later on, a jeep ripped the other one and it, too, deflated into a massive heap of canvas. Both were sent home in crates and that ended the hot air balloon offensive at Port Lyautey.

Well, that's the end of the stories I can recall right off. I would like to say that being associated with all the personnel of VP-63 was a great experience and I'm especially indebted to the great aviators and crews that had the nerve to get into an airplane with me. There was fun, tragedy, excitement, and close calls; but, I'd like to do it again. How about you?

* * * * *

MAD Cats

Editor's Note: While war is marked by combat, hardship, and heroism, it also brings out that unique "GI Spirit" these stories show – in spades. That unique ability to see the world is proof of the age-old military adage, "when you face a situation that sucks, remember suck builds character."

This bunch is not unique. Throughout the ages, military men and women have always found a unique and sometimes hilarious way of looking at their heroism. We thank them for sharing.

The Story of VP-63

A LITTLE BIT OF HISTORY
submitted by
Joseph A. Fahrner

In July the United States Navy got into the Biscay show. At Slessor's request and to provide relief for Army *Liberators* in Britain, Admiral Cane sent over Patron 63, *Catalina* flying boats, in the last week of July.

This Squadron, under Lt. Commander Edwin O. Wagner, was equipped with the Navy's new Magnetic Airborne Detector (MAD), and hence adopted the name MAD Cats. Based at Pembroke Dock in South Wales, Patron 63 began to patrol the Bay on 25 July. Two *Catalinas* flushed U-262 and U-760 about 150 miles north-west of Finisterre on the 28th and engaged them at median range while awaiting reinforcements, but the boats escaped.

The MAD proved to be of little or no use in the Bay, but one of the MAD Cats played an important, if minor, part in one of the fiercest battles in this campaign. Until late August, U-boats in transit were given very little protection by the Luftwaffe, whose capabilities with several bases near the coast of France were very great. The reason apparently was the feud between Goering and Doenitz. The portly *Reichsmarshall* did not care to build up the Gross Admiral's prestige at his own expense.

The very thin fighter and bomber control that he put over the Bay of Biscay accounted for only one victim, a MAD Cat. This was a *Catalina* named *Aunt Minnie*, piloted by Lt. William P. Tanner, USNR, who as the pilot of a PBY patrolling off Pearl Harbor, sighted a Japanese midget submarine on the morning of 7 December, 1941. Unfortunately, *Aunt Minnie* was attacked by 8 to 12 JU-88s on August 1st. She splashed one on the first pass, then came under a fatal cross fire, burst into flames from wing to wing and had to be ditched.

Tanner and his co-pilot and waist gunner climbed aboard a life raft and were rescued after 24 hours by His Majesty's ship *Bideford*. Although the Germans claimed that the MAD Cats had been chased away, they stuck to it until the end of 1943; and by that time had carried out more

MAD Cats

Bay patrols than any other squadron in Air Vice Marshall Bromet's 19th Group. [1]

Shortly after, Commander Wagner's MAD Cat Squadron, rendered ineffective by German fighter plane activities, was transferred to Port Lyautey where it (provided) excellent service patrolling the Strait of Gibraltar. [2]

Sinkings of German, Italian, and Japanese submarines by the United States forces in the Arctic, Atlantic, and the Mediterranean. [3]

The U-392 was detected by VP-63, attacked successfully by VP-63, His Majesty's ship *Affleck* and His Majesty's ship *Vanoc*. The submarine was sunk in the Strait of Gibraltar. [3]

On the 24th of February, 1944, VP-63 detected and attacked the U-761. They were assisted in the kill by VP-127, the RAF, His Majesty's ships *Anthony* and *Wishart*. This U-boat was sunk in the Straights of Gibraltar. [3] (*See Hutch's preface for the facts*)

Excerpts from *The Atlantic Battle 1 May 1943 to May 1945* by Samuel Eliot Morrison. [1] page 94; [2] page 103; [3] page 372.

October 21, 1942.

In tests with MAD gear (Magnetic Airborne Detector), a PBY from the Naval Air Station, Quonset Point, Rhode Island, located the submarine S-48. The tests were carried out in cooperation with the National Defense Research Committee. [1].

June 10, 1942.

A formal organization, *Project Sail*, was established at NAS Quonset Point for airborne testing and associated work on Magnetic Airborne Detectors (MAD gear). This device was being developed to detect submarines by the change that they induced in the earths magnetic field. Principal efforts were being carried out by the Naval Ordnance Laboratory and the National Defense Research Committee. In view of the promising results of early trials made with airships and an Army B-18, 200 sets of MAD gear were procured. [2].

The Story of VP-63

July 3, 1942.

In the first successful firing of an American rocket from a plane in flight, LCDR J.H. Hean, Gunnery Officer of Transition Squadron, Pacific Fleet, fired a retro-rocket from a PBY-5A in flight at Goldstone Lake, California. The rocket, designed to be fired aft with a velocity equal to the forward velocity of the airplane, falls vertically. It was designed at the California Institute of Technology. Following successful tests, the retro-rocket became a weapon complementary to the magnetic airborne detector with Patrol Squadron VP-63 receiving the first service installation in February, 1943. [3]

January 10, 1943.

Fleet Air Wing 15 headquarters was transferred from Norfolk to Port Lyautey, French Morocco, to direct patrol plane operations in the Mediterranean and Gibraltar Strait area. [4]

July 23, 1943.

Patrol Squadron VP-63, the first US Navy Squadron to operate from Great Britain in World War II, arrived at Pembroke Dock, Wales, to assist in the submarine patrol of the Bay of Biscay. [5]

January 21, 1943.

Headquarters of Fleet Air Wing 7 was established at Plymouth, England, to direct patrol plane operations against submarines in the Bay of Biscay, the English Channel, and the southwest approaches to England. [6]

January 18, 1944.

Catalinas of VP-63, based at Port Lyautey, began barrier patrols of the Strait of Gibraltar and its approaches with Magnetic Airborne Detection (MAD) gear, and effectively closed the Strait to enemy U-boats during day light hours until the end of the war. [7]

February 24, 1944.

The first detection of a submerged enemy submarine by the use of MAD gear was made by *Catalinas* of VP-63 on a MAD barrier patrol of the approaches to the Strait of Gibraltar. hey attacked the

MAD Cats

U-761 with retro-rockets and with the assistance of two ships and aircraft from two other squadrons, sank it. [8]

Excerpts from: *United States Naval Aviation 1910-1970.* [1] pg. 108; [2] pg. 113; [3] pg. 114; [4] pg.118; [5&6] pg. 122; [7] pg. 125; [8] pg. 127.

The Story of VP-63

MY RETURN TO ROYAL AIR FORCE BASE PEMBROKE DOCK, WALES
By Melvin Boyd Cummins

After an absence of forty-one years, I returned to the Royal Air Force Base where I was stationed with a US Navy sea plane squadron in 1943. This was a period of time when World War II was not going well for the Allied side. We were flying anti-submarine flights off the coast of France in the Bay of Biscay from the RAF base. The German Air Force also patrolled the Bay with perhaps the most versatile aircraft of World War II, Ju-88s. It was an exciting but dangerous time for the squadron.

The old sea plane base is quiet now. The giant sea planes are gone as are the flight and maintenance personnel. There is little to remind one of its active operation during World War II or the bomb damage made by the German Air Force.

The hangars are now used for ship and boat building along with heavy industry. The long sea plane ramp, extending into the channel, is now used to launch ships and boats. The tall tower-like boat loading platform dock, necessary because of the extensive tides, is no longer there along side the ramp. From this dock flight crews and maintenance men made frequent boat trips out and back to the seaplanes moored to buoys in Milford Haven Channel.

The two remaining stone headquarters buildings are on the United Kingdom's register as historical sites. They will not be demolished. Several of the barracks, that were home to the many excellent airmen and sailors, are now flats or apartments. The Officers Mess and Quarters are in a state of ruin. The roof and rooms are no longer there in most of the buildings. The front entrance to the main building still stands proudly to welcome the former distinguished aviators, if they should return. The chapel building stands sadly alone. It was used as a museum for a number of years. Now it's for sale.

The former Sergeants Mess, next to the main entrance to the base, is now the Commodore Hotel and Restaurant. The main bar and dart

MAD Cats

room seems to have retained the prestige of the "RAF Flight Sergeant" of old. The hotel provides a service for ferry passengers who depart from the near by ferry terminal for Ireland.

The Pembroke/Milford Haven area has been developed as one of the largest oil and refinery ports of Europe. The economy seems to be very good. The area is crowded with cars and people. There is hustle and bustle in the market places and shops. The streets are crowded with activity everywhere.

As I look and tour around, it seems I have forgotten most everything about Pembroke Dock except the hangars, ramp, channel, and barracks. During the short time the Squadron was based there, liberty shore leave or off duty time was not frequent. Not knowing my way around, Jane, my wife, and I were most fortunate to have been given a guided tour of the old base and all Pembrokeshire by John Evans, the aircraft researcher.

Just before departing Pembroke Dock, I walked to the to of a hill near the new Cleddau Bridge Hotel. The view to the west and north of Milford Haven Channel is outstanding. The top of the old hangars can just be seen. The channel is crowded with oil tankers and boats. Smoke stacks and oil refineries fill the skyline along with many new homes and other buildings. I find it most difficult to look back and see a PBY *Catalina* flying boat, on the step, making a wide wake in the center of the channel, with full power about to become airborne. What a beautiful view that was.

I enjoyed seeing Pembroke Dock again, the way it is now, and also touring London, all of the United Kingdom, Ireland, and Paris during this peaceful time.

Melvin Boyd Cummins and Jayne

Anacortes, Washington

The Story of VP-63

Stained Glass Memorial Window
Old Dockyard Chapel, Pembroke Dock

This window was placed in the Chapel in 1945 and remained there until after the RAF left Pembroke Dock. It was removed in 1958

MAD Cats

and placed in the Officer's Mess at RAF Mountbatten, Plymouth, another old Coastal Command flying-boat base. Although obviously in good hands and place, it is sad that Pembroke Dock's "own" window is no longer here. The local parish church, to my mind, would have been an excellent alternative location for the window.

Looking at the picture, I believe that the VP-63 badge is in the bottom right hand corner.

Excerpted from a letter from John Eva to Rear Admiral Curtis H. Hutchings 24 June, 1983. (John Evans of Aviation And Maritime Research, 6 Laws St. Pembroke Dock, Pembrokeshire, SA72DL, West Wales, UK.

The Story of VP-63

THE GUNNER RECALLS
Joseph A. Fahrner

The Reunion at San Diego was great and, during the trip back home, many incidents were brought to mind that were not part of the official history of the Squadron.

Do you recall when the Chiefs collected $300.00 and sent one of the Chiefs that specialized in "big dealing", out to an American supply ship where he dealt with his counterpart? The $300.00 was quickly exchanged for basic American food which was devoured in three days. Never knew that white bread and butter could taste so good.

Those of you who were lucky enough to get to Rabat will remember the horse-drawn taxis. There was one Arab who continuously plied his whip to the animal while chanting, "Heeb, Heeb Coco seven eleven S--O--B---Move on". That was the total extent of his knowledge of English; however, his knowledge of American money was fantastic.

And then there was Jock the Barber.

Going to NAAFI one day to make a purchase of "still lemonade", I spied some biscuits which, after purchase and eating, were promptly called "Gut Buster". I think they were made of flour, saw dust, and Elmer's glue. My chance remark was over-heard by the operators of the NAAFI which resulted in an "audience" with Captain Wagner, who agreed with me but suggested, he never ordered, I go to the Base Commander and tell him how sorry I was for my untimely remarks. I did and listened to him tell me what a hard time he had and the difficulty they had obtaining food. "Best we had, you know, carry on," he drawled.

Over-heard in London while on leave, "My God, she's got on a paper bra."

The old timers will remember "Painless Parker", dentist in the Fleet, who operated a chain of dental offices wherever there was a home port. One of our Chiefs, who later became a jg., went to Dental Shack while in Norfolk. There, he saw a Naval Dentist hold a chisel to a

MAD Cats

man's tooth and told his WAVE assistant to hit it, and she did. Our jg. promptly visited "Painless Parker" for his dental work.

While in Iceland, our Ordinance crew decided that a Mk5 floatlight would operate under water, and could become a new weapon for submarines. Our hardy experimenters went some distance from the Camp and put their device in the water. After many attempts, it ignited and went out to sea for about 20 feet, circled back to the shore, and ignited a huge pile of driftwood and wreckage. Of course, they got the hell out of there and kept their mouths shut. Grandfather of the Polaris?

Heard at a Naval establishment, "Now hear this, all those who have one, draw one. All those who have two, turn one in"????????

Then there was the enlisted man who, after taking aboard a cargo of special fruit juice, "borrowed" an officers hat and went from Nielsen hut to Nielsen hut making various types of inspection, as he put it, as we had in Kanoehe. He never went beyond Alameda. The hat was found the next day in a pile of empty beer cans, but no officer was under it. Sorry, no names.

It is axiomatic in the Navy that any chores which were not defined in "Rocks and Shoals" (Navy Regs) usually were assigned to the bosuns or gunners mates or their progeny, such as AOM's. Thus, the chore of picking up four or five '63 drunks was assigned to the Gunner and his driver. Oh yes, dear hearts, the lads did drink on occasion.

Our alcoholic shipmates were being held by the MP's at their Headquarters in Swansea, Wales. So we collected a sorry looking bunch of gobs from the Army without any paper work, a feat in itself, and threw them in the back of a 4x4 truck and covered them with a tarpaulin. These lads were so drunk that they could only sober up, get sick or both, and they did.

The 40 to 50 mile run back to Pembroke Dock did in fact semi-sober them up to the extent that we heard them calling "----Call", the first word being obliterated by the wind and the tarp. We, finally, were able to decipher the first word, they wished to relieve themselves. A lonely spot in the road was found and we stopped, thinking they would go to the side of the road in the bushes. But no, they had to

The Story of VP-63

line themselves in a semi-straight line facing the center of the road. In the midst of this natural process I saw an olive drab car approaching us. An ordinary staff car? Oh no, this one had a flag on the one fender and that flag was one star, some sort of General. What to do when you're in charge? Why you come to attention and salute.

What I did in my mind's eye I could hear my old black shoe Navy skipper telling a petty officer after Captains mast "Walk aft Seaman". But no, the driver of the staff car merely gave us a "beep, beep" on his horn and proceeded as if this was a daily occurrence.

Needless to say, we made it back to PD and got our cargo through the Main Gate by bribing the sentry with a can of American orange juice which was purloined from flight rations. Officially, the whole incident never took place as the Navy always took care of its own. The Gunner was later observed plucking several gray hairs from his, otherwise, brown hair.

Belted .50 ready for loading. Source: U.S. Army

Belting .50 caliber ammunition while in the UK was a time consuming hand-task. We had an electric belting machine, but, unfortunately, British power was not the same as in the U.S. The problem was studied and a solution was found. We needed a small gasoline engine. The nearest one that we knew of was located in Iceland on a defunct heater. It was necessary that one of our planes had to fly back to Iceland on some sort of mission and return.

MAD Cats

The ordnance man was given instructions to bring that machine back without the formality of paper work. This was done and a power transmission system was made of several rifle slings. We were in business belting .50 caliber ammunition. This operation was set up in a corner of a bombed out building as it was quite noisy and the exhaust fumes were so bad a man could only work there for a few minutes before he had to be relieved. We belted ammunition to such an extent that there never was a shortage for our planes. American ingenuity.

The Story of VP-63

FLIGHT LOG

Sniffer Wagner Spears, Mascot 1c
VP-63

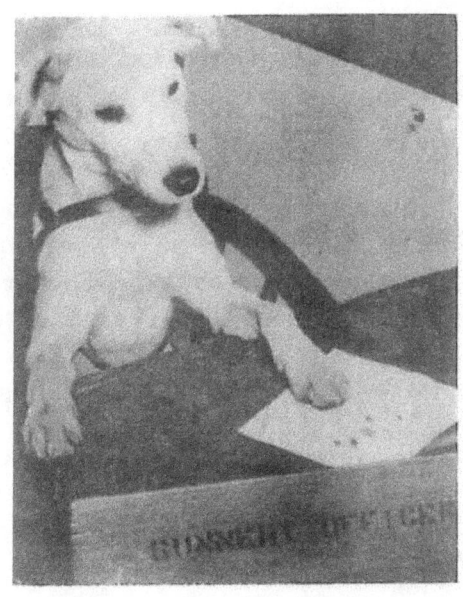

MAD Cats

This is one of the only sources available with these kind of flight details; aircraft, hours, mission.

PREFACE
The Log Of Sniffer

This is a true story of the life of a dog. The underlined entries throughout the story — the dates of the flights, type of aircraft, bureau numbers of the aircraft, number of hours and tenths of hours, and vicinity of flights — are taken directly from the Flight Log of Sniffer. The dog belonged to CDR R. C. Spears, US Navy, and since Sniffer's retirement in Clarendon, Arkansas, his owners are Mr. and Mrs. R. L. Pace of that City.

Ensign R. C. Spears, the owner of Sniffer was stationed at Kaneohe, Hawaiian Islands and in the air when the Japanese attacked Pearl Harbor on December 7, 1941. For at least two weeks after that day of infamy, it was bedlam at the base — no aircraft to fly, black outs, rumors of invasion, etc. Ensign Spears decided he needed a dog to sleep on his bed and alert him of danger while sleeping. In 1942 Ensign Spears returned to Alameda, California, for the formation of Patrol Squadron Sixty-Three. Sniffer was purchased at a pet shop in San Francisco and trained to be a watch dog. Sniffer was the name used in VP 63 for the MAD gear (Magnetic Anomaly Detection) that was developed by Columbia University and successfully used by VP 63.

Sniffer was a charter member of the squadron. He retired from the Navy in 1945 and spent the remaining years of his life with Mr. and Mrs. R. L. Pace, Clarendon, Arkansas. Mrs. Pace was the "Aunt Minnie" for whom Ensign Spears' plane was named. It was shot down by enemy Ju 88s in the Bay of Biscay in 1943. Lt. Tanner and his crew were flying *Aunt Minnie* at the time.

Sniffer died in 1955.

The Story of VP-63

The Log of Sniffer

I was born a dog sometime around the middle of January 1943 in the bay city of San Francisco, California. There seemed to be lots of unusual activity in those days, people in uniform were everywhere. My first recollection was in a show window on Market Street in the city of my birth when a LT(jg), U.S. Navy, came into the store with a little girl looking for a rabbit. He had a ten dollar bill and the new sales girl had been briefed well by the boss for she went all out to get the whole ten, she offered me as one of the things, other than the rabbit, that could go for ten dollars. So out of the store we go, I'm under one arm, the rabbit under the other, one hand was clutching the little girl, and the darn fool had a big package of rabbit food. He was loaded down. We got into the car and out to Alameda we go. The BOQ home there was nice, but it was lonesome for me. My new owner didn't come in until late at night, and he would get up early in the morning and take off for somewhere, the Lord only knows. We lived that way for about five weeks and then one day he took me to the office with him and we go flying in a PBY airplane. What an experience, the noise was unbearable, but my master seemed to enjoy my being with him and the food and attention wasn't bad.

<u>14 March 1943   PBY 5 08231 1.0 hour Local</u>

The next thing I know, he is calling me Sniffer and my full name was Sniffer Wagner Spears. Now that name seemed strange to me; but since then, I have learned that I was named after the gear that the planes were carrying, which was a big Secret. It was supposed to smell submarines. The Wagner was after the Commanding Officer of the Squadron, and of course, my last name was after my master. After flying in those darn contraptions for some few days, my name didn't mean much anyway, Hell, I couldn't hear or smell half the time.

<u>14 March 1943   PBY 5 08231 3.4 hours Alameda to San Diego</u>

We departed for the East coast via San Diego in March and it was some trip. I caught a bad case of GI's at the Salton Sea and my

MAD Cats

master fed me cheese like I was a member of a Chamber of Commerce in Wisconsin.

16 March 1943   PBY 5 08231 1.0 hour San Diego to Salton Sea

I even overheard him talking to some sharp cookie, another Navy jg, and he said a cork stopper might work.

17 March 1943   PBY 5 08231 9.0 hours Salton Sea to Corpus Christi

I got lots of sleep and good chow; but it wasn't long and we take off for Corpus Christi, Texas, Now there was a city! We moved into a big hotel, the tallest one there. A shorter one sits right next to it. There are various stories about these two hotels and I'm sure that you have heard about them. My master would bring me food, but the walks and great outdoors jaunts for me stopped and the only place for me to go was the rug. He seemed mad every time he came in at night about cleaning up after me; but I feel sure that he realized that it wasn't all my fault. After the third day there, the manager of the hotel came up and asked my master and me to please vacate. We left there and moved to the BOQ on the Naval Air Station.

20 March 1943   PBY 5 08231 6.5 hours Corpus Christi to Pensacola

We departed that lovely Texas city on 20 March and flew to Pensacola, Florida. Can't seem to remember much about that place except the facilities were about the same as all the other stations.

22 March 1943   PBY 5 08231 3.2 hours Pensacola to Jacksonville

We next went to Jacksonville, Florida. It was a spitting image of Pensacola as far as I was concerned.

23 March 1943   PBY 5 08231 5.4 hours Jacksonville to Elizabeth City

On 23 March we take off for Elizabeth City, North Carolina. Here we settled down for about ten days. I did lots of flying on routine patrols and test hops; also my master made me a full member of his crew.

The Story of VP-63

<u>28 March 1943   PBY 5 08231 2.5 hours Routine hop</u>

Dear old North Carolina is the place where I learned, after much coaxing, that I was supposed to go outside, this was a relief to all concerned.

<u>29 March 1943   PBY 5 08231 2.5 hours Routine hop</u>

All the planes, some 15 of them had sniffer gear, but I don't remember any of them sniffing a darn thing. I was getting used to the corn flakes, sugar, and milk that my master usually fed me for meals. Also, they have a canned meat with a four letter spelling that I first tried here, rhymes with ham. Deliver me from it!

<u>30 March 1943   PBY 5 08231 3.0 hours Night local</u>

<u>2 April 1943   PBY 5 08231 5.0 hours Local Patrol</u>

<u>6 April 1943   PBY 5 08231 4.0 hours Elizabeth City to Quonset Point</u>

On 6 April we take off for Quonset Point, Rhode Island. Gee it was cold there in April, I nearly froze to death, but we finally got beached and checked into a room in the BOQ. Here I got fixed up with an ID card, picture and all. My master had a parachute made for me with a harness and I overheard him talking about me going to jump out of the plane at 500 feet over the station. He weighed me and had the dern harness fitted; but I never made the jump because permission was not granted. It seems that permission was requested from the Commanding Officer of Quonset and the SPCA. The Commanding Officer approved but the SPCA did not, so we carried the thing and I posed and strutted a few times for onlookers.

<u>8 April 1943   PBY 5 08231 2.0 hours Familiarization hop</u>

<u>15 April 1943   PBY 5 08157 11.7 hours Operational patrol</u>

<u>25 April 1943   PBY 5 08157 9.0 hours Operational patrol</u>

<u>29 April 1943   PBY 5 08157 2.0 hours Local hop</u>

May was a big flying month for me. Some commercial airline pilot spotted nine German submarines on the surface just off Bermuda, and Wagner's men with the Sniffer gear was called to the rescue.

MAD Cats

We sniffed the whole area around that lovely island, but it was to no avail. It seems that the enemy always heard about us coming or by coincidence they were never in our path.

3 May 1943   PBY 08146 8.4 hours Quonset to Bermuda

4 May 1943   PBY 08146 10.0 hours Submarine Patrol

6 May 1943   PBY 08146 11.2 hours Submarine Patrol

8 May 1943   PBY 08146 11.1 hours Submarine Patrol

11 May 1943   PBY 08174 11.1 hours Submarine Patrol

13 May 1943   PBY 08176 11.2 hours Submarine Patrol

16 May 1943   PBY 08176 11.0 hours Submarine Patrol

18 May 1943   PBY 08176 15.0 hours Night Convoy Coverage

23 May 1943   PBY 08176 3.8 hours Searching for Survivors

We departed Bermuda for Quonset on the 23rd of May and we all sort of rested and got our planes back in good shape from all the flying we did in Bermuda. We received a commendation from Commander Bermuda for a night hop where we found the life rafts full of personnel of a U.S. Navy downed *Coronado*. It was a great day for the Old Man.

23 May 1943   PBY 08176 8.0 hours Bermuda to Quonset Point

We spent almost a month in Quonset with routine duties, Made a trip to Boston and I became lonesome as the old man kept going out a lot without me.

22 June 1943   PBY 08231 7.3 hours Quonset to Argentia, Newfoundland

I was anxious to get back into the game of flying, so on 22 June 1943 we take off for Iceland via Argentia.

26 June 1943 PBY 08231 15.7 hours Argentia to Iceland

It was nice weather in Iceland, but I could never tell when to go to bed and my master didn't know either. The sun was shining 24 hours a day. Now here I met lots of friends, but I admit I was wondering whether everyone else was shaped different or was I just

The Story of VP-63

different. Since, I have seen other dogs; but for a while I wondered. We flew and flew and again no submarines and no chance to prove the sniffer gear. Again the enemy must of heard of our coming. Anyway, no one told me what I was supposed to do so I just lay in the bunk of the plane on most of the hops. Pretty good duty I'd say.

<u>16 July 1943   PBY 08231   4 hours Arctic Circle</u>

On 16 July, I make a routine hop north across the Arctic Circle and a Blue Nose certificate was issued to me, but for what I'll never know. With that weather my nose was blue most of the time anyway.

<u>22 July 1943   PBY 08231   10.8 hours Iceland to Pembroke Dock</u>

On 22 July, we takeoff for Pembroke Dock, Wales. When we arrived there, a civilian dog man was waiting for me. He told my master that dogs would have to be sent to London for six months quarantine. Now there I was, a full-fledged flying dog, doing my best, plane broken, and they accused me of having rabies. Well my boss asked the guy if we could stay on the RAF base if he constructed a special enclosure. He didn't know, but said he would write London and find out. The enclosure was built and I moved into it but not for long, as the man came back and advised London had said NO.

To London I must go or return to the United States. My master chose the states as some friends of his were leaving by train for London and then by boat to the United States. I made the trip to London; but when we got there these so-called friends left me at the Fleet Post Office with a sailor and there I lived for about a month until one day a Lieutenant came in and explained to the sailor that I lived in Pembroke Dock and he was to take me home to my master.

 I never did really know how he worked it, except we never saw that dog man again and my master kept telling the local people in Wales that I was born in England, and wasn't I odd for an English dog.

I received my First Class Mascot designation in September and my flying time was curtailed.

MAD Cats

PATROL SQUADRON SIXTY-THREE SEPTEMBER 2, 1943, SNIFFER WAGNER SPEARS HAS THIS DATE QUALIFIED AS A MASCOT 1/c, IN ACCORDANCE WITH ANY SECTION, AIRCRAFT SCOUTING FORCE MANUAL.

E.O. Wagner Lieut. Comdr. USN. Commander, Patrol Squadron 63

12 September 1943   PBY 08174 9.2 hours Bay of Biscay Patrol

Now let me tell you about these English dogs. They are famous and I liked them, especially the female variety. There were rumors that I was a little out of my class; but with that nickname and my flying ability, I had no trouble with the opposite sex. My flying here was in the Bay Biscay and we sniffed the whole bay, but again the enemy must have heard about us coming with all the new devices.

I made trips to Hamworthy, England, the Scilly Islands, Mt. Batten, and let me tell you my master was always flying in the soup. If there were clouds, he would fly in them, if none he was looking for them. It seems that the Germans had some fighter planes over the area and they were on the lookout for a slow flying boat.

24 September 1943   PBY 08145 7.2 hours Operational Patrol

I only flew in a few of the hops, but my master was out flying every third day. I loved England and from my point of view, there will always be an England. Even today, I dream of England and my youth.

2 October 1943   PBY 08145 12.3 hours Operational Patrol

29 October 1943   PBY 08439 11.7 hours Patrol, Scilly Islands

18 December 1943   PBY 08145 12.0 hours Pembroke to Pt. Lyautey

I left Pembroke Dock on 18 December 1943 on a long hop bound for Port Lyautey, French Morocco. Had lots of company on the mass flight as many of my masters shipmates had purchased dogs in Wales. There was Radar, a cocker spaniel; Pembroke, a little white fuzzy dog; Buddy, a field spaniel; and a whole host of other dogs that went along. When we hit the beach in Port Lyautey there was

The Story of VP-63

eleven of us dogs and things were looking up toward a nice tour. Here my master started giving me lots of advice and training routine.

<u>8 January 1944   PBY 08145 9.5 hours Operational Patrol</u>

I made a few hops, but most of my time was spent in learning new ground procedures. Hour after hour we practiced to make me a watch dog. My master would lay down on the bunk and I would stay on the foot of the bed and was supposed to bark when anyone came near the bed. I became pretty good, at least everyone said that I did; but all I did was growl and look mean. Don't think I ever had the nerve to bite anyone. I was also trained on the words "Jap" and "German". When these words were spoken, I was supposed to bark and look very ferocious, which I did.

I learned to sit up, play dead, roll over, and I guess my greatest triumph was being house broken. You see, most of my dog buddies just didn't have the word. Most of us lived together in one Quonset hut and every night there was lots of activity in the huts and my buddies would have to go and the deck was handy as both doors were closed. Well, every morning the officers would start accusing each other about his dog, my dog, and so on. Every master was accused. It seems all dog owners were trying to match the evidence with a dog, and that is some problem I will say.

It got pretty bad and my master became sore, so he said he would put me in the bed with him and someone could tuck the mosquito net down all around the bed. If it was not moved during the night and I was still in the bed come morning, my master and I would not be guilty. Well this worked for a long time until one night I was awakened and there was this Lieutenant, one of my bosses better buddies, cutting a hole in my master's net.

He told me to be quiet, so keep quiet I did. Anyway the next morning there were three patches of evidence and I was blamed for the whole lot. Oh, the life of a dog. There were sharp words by all concerned and this is one time that I guess I should have talked. My master then pitched in and helped the rest of the officers train their dern dogs.

MAD Cats

13 January 1944   PBY 08145 9.5 hours Operational Patrol

28 January 1944   PBY 08146 5.4 hours Anti-Submarine Patrol

8 February 1944   PBY 08174 7.6 hours A/S Patrol Gibralter

9 February 1944   PBY 08174 7.7 hours A/S Patrol Gibralter

15 March 1944   PBY 08154 6.8 hours A/S Patrol Gibralter

16 March 1944   PBY 08154 7.3 hours Sniffed Sub — Sank Same

In March 1944, we sniffed the jackpot after all the many months we were still trying to find something so we could prove the Sniffer gear. At 0755 on 16 March in the Straits of Gibraltar, sniffed sub. That was the day. Sank same. Got a definite kill, too. The German sub never knew what tracked or hit him. I will always remember those explosions.

I was proud of my name on that day and nothing was too much for my master. I was presented an ice cream cone and a chocolate Hersey bar when we returned to base. These are still two of my favorite dishes. Port Lyautey was nice, but the City of Rabat, about 15 miles away, was better. Have you ever heard about those French dogs? Well, if you haven't, it is just as well.

By August, 1944, I had built up a grand total of 441.8 flying hours and our old plane was badly in need of overhaul, so on the 19th of August we take off for the states.

19-20 August 1944   PBY 08154 56.8 hours Port Lyautey – Dakar – Natal – Belem – Trinidad - San Juan – Jacksonville – Norfolk

I rode my first train from Norfolk, Virginia to Clarendon, Arkansas. It took us about seven days to make this trip. For a while, I thought the end of my time had come. My master left me in a kennel in Norfolk for shipment and he went on home earlier to marry some Arkansas lady.

On 9 September, 1944, my master got married and I was a right sick dog due to my rough trip on the train from Norfolk to Clarendon. The new lady of the house didn't like me very much and she was continually scolding my master for the first ten days of his married life because he was showing me so much attention. Three's a

The Story of VP-63

crowd. I had been taught to bark and growl when anyone got in the bed with my master and I always gave her a small growl — wasn't serious though; but it got a rise out of the old girl.

We had always had wonderful times together and with the addition of the new lady things changed, but not to my satisfaction. Anyway, the old girl would feed me occasionally and, in time, she learned my name; but I doubted if we would ever be able to mutually understand each other. We moved to Quonset Point, Rhode Island, and I became a routine, average, everyday, house dog. I would go to the hangar occasionally with my master, but my flying days were few.

<u>19 November 1944 PV-1 33143 1 hour Local</u>

<u>25 November 1944   PBY 5A 08101 1.0 hour Quonset – Local</u>

<u>24 April 1945 PBY 5A 48421 1.0 hour Local</u>

On July 7, 1945, a little baby came to our house and I liked him; but Mrs. Master kept insinuating that I might bite him, and I could see the handwriting on the wall. My attentive master was losing attentiveness to me. I lost my pride and people spoke of me as average just like all the rest of the dogs in the neighborhood.

In September, 1945, we moved back to Arkansas and my master became a coach at Dermott High School. This was fine for me, because I would go to classes, lots of kids would feed me candy, cookies, and ice cream. I performed on the stage for the grade school students, and this restored a lot of my pride. I guess the year I spent in Dermott was one of the happiest years of my life.

In July, 1946, my master was transferred back into the Navy and he departed for Guam. Well, I moved to Clarendon with his folks and sort of retired.

I have been living here ever since and this Arkansas is a nice place to retire and lead a normal dog's life. I like my new masters; they feed me well and give me all kinds of attention. I weigh some 65 pounds now against 30 when I was in my prime, but you know — middle age spread.

MAD Cats

Appendices

MAD Cats

The Story of VP-63

Appendix A

CHRONOLOGY   July 23, 1993

Naval Aviation History section

9-19-42 Commissioned, NAS Alameda
         Aircraft squadron desig., CNO serial 042531 of 17 Feb 45
10-1-44 Changed to VPB-63
7-2-45   Decommissioned Norfolk, VA

Commanding Officers
9-19-42 Lt Cdr E.O. Wagner
9-23-43 Lt Cdr Curtis H. Hutchings
7-25-44 Lt Cdr Carl W. Brown

2 Nov 1942	Ferry trip by 9 PBY-5As from San Diego, CA to Kanoehe Bay, HI
5-6-7 Dec	Carried out search patrols in Pacific for Japanese carriers believed to be coming in under cover of a front to make a sneak raid on California coast.
15 Amrch 1943	Departed Alameda for San Diego
16 March	Det from FAW-8 to FAE-5
17 March	Departed San Diego for Salton Sea
19 March	Departed Salton Sea for Corpus Christi, TX
21 March	Departed Corpus Christi for Pensacola
22 March	Departed Pensacola for Jacksonville
23 March	Departed Jacksonville for Elizabeth City, NC where based temporarily for searchlight training
4 April	6 planes detached for temporary duty at Key West (reported to FAW-12)
6 April	Departed Elizabeteh City for Quonset Pt. for operation under ComAsDevLant
11 April	Admin transferred from FAW-5 to FAW-9
24 April	4 planes detached for temporary duty at Jacksonville
2 May	Seven planes to Bermuda for A/S operation
24 May	Planes of Bermuda detachment returned to Quonset Point
6 Jun	Jacksonville detachment returned to Quonset Point
22 Jun	VP-63 departed Quonset Point for Argentia, Newfoundland (Admin transferred to FAW-7)
26 Jun	VP-63 departed Argentia for Reykjavik, Iceland for operation under FAW-7
20 July	Departed Reykjavik for Pembroke Dock, South Wales, for operation with RAF Coastal Command under FAW-7. Was first

MAD Cats

	US squadron to operate from United Kingdom in campaign against enemy U-boats.
26 Dec	Admin transferred from FAW-7 to FAW-15
20 Jan 1944	2 planes left in England joined squadron
25 July	Lt Crd Carl Wallace Brown relieved Cdr Hutchings
	Operations under FAW-15 consisted of anti-sub warfare in Gibralter & Mediterranean area. Use of MAD. equipment

VPB-63
1 Oct 1944	Designation of squadron changed to VPB-63
	VPB-63 Secret ltr (VPB-63/A12-1/Wal) of 11 Dec 44
Oct-Dec 1944	FAW-15 PBY-5 Port Lyautey, French Morocco. A/S patrols in Straits of Gibralter & Mediterranian area
6 Dec	Squadron reduced to 12 aircraft with 3 spares
11 Dec	Surplus personnel returned to US
	VPB-63 sec ltr, ser 001-45 of 20 Apr 45, Squadron history
Jan 45	FAW-15 PBY-5 Based at NAS Port Lyautey, French Morocco; operational flights
10	FAW-7 PBY-5A Morocco; operational flights
Feb 45	FAW-15 Based NAS Port Lyautney, French Morocco Detachment based at Dunkeswell, Devon
Mar 45	Same
Apr 45	Same
May 45	FAW-15 NAS Port Lyautney, French Morocco
	TF(R) 125 Part of Detachment at Dunkeswell, Devon ret (planes)
June 4	Rec'd orders to proceed to Norfolk for decommissioning
June 21	FAW-5 All personel arrived Norfolk
June 22	Decommissioning date set on or before 2 July
July 2	Decommissioned at Norfolk

W.D. VPB-63 for Jan 45 serial 05-45 of 1 Feb 45
W.D. VPB-63 for Feb 45 serial 07-45, film 108067 of 1 Mar 45
W.D. VPB-63 for Apr 45 serial 013-45 of 1 May 45, film 118710
W.D. VPB-63 for May 45, serial 017-45 of 1 June 45, film 122912
VPB-63 sec ltr, ser 002-45 of 23 June 45, Squadron history
VPB-63 sec ltr, ser 001-45 of 20 Apr 45, Squadron history

The Story of VP-63

Appendix B

PATROL BOMBING SQUADRON SIXTY-THREE
c/o Fleet Post Office, New York, New York

From: Commander Patrol Bombing Squadron SIXTY-THREE.
To: History Unit, Op-33-J-6, Office of Editorial Research, Aviation Training Division, Office of the Chief of Naval Operations.
(1) Commander Fleet Air Wing FIFTEEN.
(2) Commander Air Force, U.S. Atlantic Fleet.

Subject: Squadron History - submission of.

Reference: (a) Aviation Circular Letter #74-44.

Enclosure: (A) "The Story of the Mad Cats", A History of Patrol Bombing Squadron SIXTY-THREE.

1. As required by reference (a), "The Story of the Mad Cats," A History of Patrol Bombing Squadron SIXTY-THREE is submitted as enclosure (A).

(Signed)
C.W. Brown

MAD Cats

PART I.

CHRONOLOGY

1942	September 19	Patrol Bombing Squadron Sixty-Three commissioned at U.S. Naval Air Station, Alameda, California, under Patrol Wing SEVEN. Lieutenant Commander E. O. Wagner, U.S.N., Commanding Officer.
	November 2	Ferry trip by nine PBY-5A's from San Diego, California, to Kanoehe Bay, T.H.
	December 5,6,7	The squadron carried out search patrols in the Pacific for Jap carriers believed to be coming in under cover of a front to make a sneak raid on the California coast.
	December 31	PBY-5A airplane (Bureau No. 08100) made a night crash landing five miles west of Heceta Head Light, Oregon. The plane sank in forty-five minutes, where-upon the crew took to life rafts. However, only one man was able to reach shore safely, and the others were lost in the heavy surf. The men lost were: Lieutenant James Edward BREEDING, A-V(N), U.S.N.R. Ensign Alvin L. CHAMBERLAIN, A-V(N), U.S.N.R. HELMING, Loyd, 321 22 88, AMM1c, U.S.N. BURTZ, Henry LeRoy, 268 86 16, AMM3c, U.S.N. O'CALLAGHAN, George Franklin, 644 32 52 ARM3c,U.S.N.R GORDON, Howard Stanley, 316 41 82, ARM1c, U.S.N.

The Story of VP-63

SMITH, Newell, SK2c, U.S.N. (Passenger)

The sole survivor was:

Ensign Roderick McCleod BRUSH, A-V(N), U.S.N.R.

1943 January 1 to March 1 Fifteen new PBY-5's received from factory and converted for M.A.D. operation.

January 21,22,23,24 All available planes conducted searches in the Pacific for a clipper lost with high ranking officers aboard. The plane was later found crashed in the mountains.

February 14 PBY-5 airplane (Bureau No. 08158) made a crash landing in San Francisco Bay, California, and the following men were killed or drowned as a result of the crash:

Lieut.(jg) Henry KOVACS, A-V(N), U.S.N.R.
JOHNSON, Jack Hubert, AP1c, U.S.N.
O'DELL, Hiram Crover, ARM3c

The remainder of the crew survived the crash landing and were able to row ashore in a life raft. They were:

MOORE, George Newton, AMM1c, U.S.N.
PEARSON, Francis Roy, AMM3c, U.S.N.
PETERSON, Richard George, ARM1c, V-3, U.S.N.R.
HOLLEN, Frederick Marcus, RT1c, V-3, U.S.N.R.
 (Passenger)
ROSE, Earle Tarleton, RT1c, V-3, U.S.N.R.
 (Passenger)

		FERGUSSON, Eugene Stewart, RT1c, V-3, U.S.N.R. (Passenger)
	March 15	VP-63 departed Alameda, California, for San Diego, California.
	March 17	VP-63 departed San Diego, California, for Salton Sea, California.
	March 19	VP-63 departed Salton Sea, California for Corpus Christi, Texas.
1943	March 21	VP-63 departed Corpus Christi, Texas, for Pensacola, Florida.
	March 22	VP-63 departed Pensacola, Florida, for Jacksonville, Florida.
	March 23	VP-63 departed Jacksonville, Florida, for Elizabeth City, North Carolina, where the squadron was temporarily based for searchlight training.
	March 30	PBY-5 airplane (Bureau No. 08251) crashed about one and a half mile off Wade Point Lighthouse, Albemarle Sound, North carolina, while on searchlight training flight. The entire crew was lost. They were: Lieutenant Frederick A. BROWN, Jr. U.S.N. (5797) Ensign Robert D. WARREN, Jr. A-V(N), U.S.N.R. (130308) GRATTON, Louis Wilson, 321 32 16, AP1c, U.S.N. NEWTON, Guy Wilson, 360 04 11, AMM1c, U.S.N.

The Story of VP-63

	BENTLEY, LaVerne Deheart, 234 13 39, AMM2c, U.S.N.
	TUROVER, Maurice Bernard, 201 86 91, ARM2c, U.S.N.
	ABRANZ, Walter Kenneth, 350 99 10, ARM3c, U.S.N.
	HARBOUR, Eugene Franklin, 636 03 17, S2c, V-6, U.S.N.R.
April 4	Six planes detached for temporary duty at Key West, Florida.
April 6	VP-63 departed Elizabeth City, North Carolina, for Quonset Point, Rhode Island, for operation under Commander Aircraft Anti-Submarine Development Detachment, Atlantic Fleet.
April 18	Key West detachment rejoined the squadron at Quonset Point, Rhode Island, via Elizabeth City, N.C.
April 24	Four planes detached temporary duty at Jacksonville, Florida.
May 2	Seven planes sent to Bermuda for anti-submarine operation.
May 9	Squadron plane No. 6, Lieutenant Kauffman patrol plane commander, landed in the open sea to rescue four survivors from a torpedoed merchant vessel.
May 10	PBY-5 airplane (Bureau No. 08226) was lost on a training and anti-submarine flight from Jacksonville, Florida. A gas line failure made it necessary to land in the open sea which was very rough. All personnel were

MAD Cats

		rescued by <u>PT 255</u> and by <u>U.S.S. Tyrer</u>, which was homed to the scene by other planes of the squadron. The latter vessel attempted to tow the plane, but this became impossible as the weather became worse and it was abandoned.
	May 23	Planes of the Bermuda Detachment, with Lieutenant Commander Hutchings and Lieutenant Spears as patrol plane commanders, assisted in the rescue of survivors from a PB2Y.
	May 24	Bermuda Detachment returned to Quonset Point, R.I.
	May 30	PBY-5 airplane (Bureau No. 08175) was lost following a crash landing in Wickford Harbor, Quonset Point, Rhode Island. The crew took to rubber boats and were quickly rescued by the crash boat from the base. There were no fatalities and only minor injuries.
	June 6	Jacksonville Detachment returned from Quonset Point, Rhode Island.
1943	June 22	VP-63 departed Quonset Point, Rhode Island, for Argentia, Newfoundland.
	June 26	VP-63 departed Argentia, Newfoundland, for Reykjavik, Iceland, for operation under Fleet Air Wing SEVEN.
	July 20	VP-63 departed Reykjavik, Iceland, for Pembroke Dock, South Wales, for operation with the R.A.F. Coastal Command under Fleet Air Wing

The Story of VP-63

SEVEN. It was the first United States Navy squadron to operate from the United Kingdom in the campaign against enemy U-boats.

August 1

Squadron Plane No. 10 (Bureau No. 08231) was shot down by eight German JU-88s in the Bay of Biscay. The following men were lost, either killed by enemy fire or in the crash landing:

Lieutenant (jg) Billy England ROBERTSON, A-V(N), U.S.N.R.
GOLDER, William Henry, 380 72 05, ACRM(AA), USNR
SCOTT, Raymond Carl, 412 09 77, ACRM(AA), V-3, U.S.N.R.
RITTEL, Arthur Alfonso, 342 02 61, AP1c, USN
RUDE, Werdin Onell, 328 94 33, ARM2c, USN
CARMACK, David Robert, 376 24 30, AMM2c, USN
LAW, Robert Bonar, 652 11 98, AMM3c, V-6, USNR

August 2

The following men survived the attack and crash landing on August 1, and after spending twenty hours in the open sea in a life raft were rescued by H.M.S. Bideford:

Lieutenant (jg) Robert Irving BEDELL, A-V(N), U.S.N.R.
Lieutenant William P. TANNER, Jr. A-V(N), USNR
PATTERSON, Douglas C., 642 76 88, ACM3c, V-6, U.S.N.R.

September 23

Commander E.O. Wagner was relieved by Lieutenant Commander Curtis H.

MAD Cats

		Hutchings, USN, as Commanding Officer of VP-63.
	December 16	VP-63 departed Pembroke Dock, South Wales, for Port Lyautey, French Morocco, for operation under Fleet Air Wing FIFTEEN. Two planes remained in England for secret experimental operation.
1944	January 9	Squadron Plane No. 11 (Bureau No. 08172) crashed during take-off in the Oued Sebou River at Port Lyautey. The following members were killed:
		Lieutenant Woodrow E. SHOLES, A-V(N), USNR
Lieutenant (jg) MArvin S. CLINTON, A-V(N), USNR
KETTMAN, John J. Chief Aviation Pilot, USN
BROWN, Walter E. ARM1c, V-3, USNR
WATERMANN, Wilhelm H., ARM2c, V-3, USNR
MANNING, James O., Jr., AMM2c, V-2, USNR
CHRISTENSON, Edwin P., AMM3c, USN
		Survivors of the crash were:
		SKELTON, James W. ACMM(AA)
KEANE, Milton F., AOM2c
	January 20	Squadron Plane No. 4, Lieutenant V.M. Mayabo, patrol plane commander, and Squadron Plane No. 15, Lieutenant R.A. Barton, patrol plane commander, arrived at Port Lyautey to rejoin the squadron after special duty in England.
	February 24	Squadron Plane No. 1 Lieutenant (jg) T.R.

The Story of VP-63

	Woolley, patrol plane commander, and Squadron Plane No. 14 Lieutenant (jg) H.J. Baker, patrol plane commander, bombed a submerged submarine while on patrol in the Strait of Gibraltar. The British destroyers <u>Wisehart</u> and <u>Anthony</u> also attacked. Forty-eight survivors, including the captain of the German U-boat were picked up by the <u>Anthony</u>, and the destroyed submarine was identified as the U-761, tentatively listed as 540 tons.
March 16	Squadron Plane No. 8, Lieutenant R.C. Spears, patrol plane commander, and squadron Plane No. 1, Lieutenant (jg) V.A.T. Lingle, patrol plane commander, attacked a submerged submarine in the Strait of Gibraltar. Squadron Plane No. 7, Lieutenant (jg) M.J. Vopatek, patrol plane commander, assisted in tracking the submarine. A British frigate, <u>H.M.S. Affleck</u>, also attacked. The Flag Officer of Gibraltar and Mediterranean Approaches evaluated the attack as resulting in the probable destruction of the submarine by combined action, and this assessment was later confirmed by the British Admiralty.
April 15	Squadron Plane No. 13, Lieutenant M.W. Nicholson, patrol plane commander, was flown on a series of tests in the Bay of Bizerte, Tunisia, to determine the

MAD Cats

practicality of using MAD gear to detect mines.

May 15   Squadron Plane No. 12, Lieutenant (jg) M.J. Vopatek, patrol plane commander, and Squadron Plane No. 1, Lieutenant H.L. Worrell, patrol plane commander, attacked a submerged submarine in the Strait of Gibraltar. Attacks were also made by H.M.S. Blackfly, a British armed trawler, and by H.M.S. Kilmarnock, a British escort craft. The attack was evaluated by the British Admiralty as resulting in the probably destruction of the submarine.

The Story of VP-63

Appendix C

NAVY DEPARTMENT

IMMEDIATE RELEASE
PRESS AND RADIO

JULY 7, 1945

PATROL BOMBING SQUADRON SIXTY-THREE

The first U.S. Naval Aviation squadron to operate from the United Kingdom in the war against German submarines has returned to the United States.

Patrol Bombing Squadron Sixty-three, credited principally for its work in developing and maintaining an impenetrable barrier that denied passage through the Straits of Gibraltar to enemy U-Boats, inaugurated daylight Navy anti-submarine patrols over Nazi fighter protected Bay of Biscay almost two years ago to the day of their return home.

This squadron is credited with officially with sinking three underseas craft during its 24 months of foreign duty and is listed as having inflicted probable damage to others. The record of achievement also includes shooting down two JUNKERS-88s in an engagement that marked the first aerial meeting between U.S. Naval Aviation and the Luftwaffe in the European Theatre of Operations.

The attack developed far out over Biscay when eight German JU-88's jumped the Navy CATALINA known affectionately to all in the squadron as "Aunt Minnie." This lumbering flying boat didn't have a chance against the twice-as-fast Nazi attackers, yet two of them were shot down before she crashed into the Bay with seven of hew ten crew already dead.

During its stay in the United Kingdom, Patrol Bombing Squadron Sixty-three flew more hours against the enemy than any other Squadron engaged in similar operations under the Royal Air Force Coastal Command Group to which it was attached. Sorties average nearly 12 hours each in duration and were conducted in weather that often proved a greater hazard than the constant threat of attack by German planes.

MAD Cats

The squadron was based in North Africa during its anti-submarine barrier patrol on the Atlantic side of the approaches to the Straits of Gibraltar. CATALINAS operating in two-plane relays, maintained dawn-to-dusk vigil over the narrow channel leading to the Straits. They flew at a dangerously low level. Often the planes had to gain altitude to avoid wing tips striking the water as tight turns were made to keep outside the three mile limits of neutral Spain and Spanish Morocco.

In the process of turning the Mediterranean Sea into an Allied lake, Patrol Bombing Squadron Sixty-Three sometimes averaged better than 1,300 hours of operational flying per month. Its Gibraltar U-Boat "fence" is credited with being a major factor in the successful invasion of Southern France since no allied vessel was sunk by enemy submarines.

A detachment of the squadron returned to the United Kingdom during the closing months of the war to participate in the attacks that marked the final flurry by the U-Boats. It was led by Lieutenant Commander Carl W. Brown, U.S.N.R., 1303 Ocean Drive, Corpus Christi, Texas, wrote who is the third skipper of the Naval Aviation Unit. After specialized training by Air Force, Atlantic Fleet, personnel of Patrol Bombing Squadron Sixty-Three will be transferred to the Pacific Theatre of War.

VP-63 De-Commissioning Ceremony. Source: U.S. Navy

Appendix D

ODE TO A PBY

Blessings on thee, you PBY
Staggering through a stormy sky,
Struts and fittings loose and worn,
And thy fabric loose and torn.

Rivets loose and flapping wings,
Leaky hulls and other things,
Your engines spit and pop like heck,
Twelve more hours 'till a check.

When the airspeed meter climbs
Past one hundred eighty-nine,
The crews hearts sink in their boots,
And they hunt their parachutes.

Bottoms torn on jagged rocks,
Wing tip floats left on docks,
Paint all gone from landings hot,
Rivets sheared by full stall squats.

Control wires slack and frames askew,
Hatches stuck and corroded screws,
Oft a wonder when on high,
Why it is you still can fly.

Everytime I fly with you,
In misty night or morning dew,
I know you must return with me,
Or be with me eternally.

John D. Linhart, AMM
(Submitted to VP-63 2nd Anniversary Book)

MAD Cats

The Story of VP-63

Appendix E

PBY-5/5A Specs

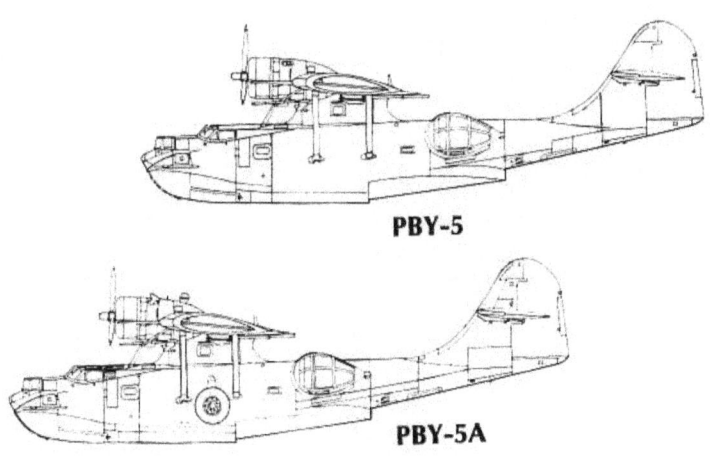

PBY-5

PBY-5A

	Gross Wt.	MPH/Alt	Service Ceiling	Patrol Range	Climb ft/min	T/O Time calm sea
PBY-5 (early)	26,200	200/5700	21,600	1965	990/1	25
PBY-5 (late)	31,813	195/7000	17,700	2860	660/1	45
PBY-5A	33,975	180/7000	14,700	2545	880/1	30

MAD Cats

The Story of VP-63

Selected Bibliography

Ashworth, Chris, *RAF Coastal Command*, Patrick Stephens Limited, Hayned Publishing Group, Yeovil, Somerset, UK, 1992.

Craddock, Lt. William P., USNR, *The History and Tales of the MAD Cats (Now It Can Be Told): By the Officers and Men Who Operated and Maintained Them*, Occasioned by the Forty-First Anniversary of Navy patrol Squadron VP-63, San Diego, California, September 15-18, 1983.

Craddock, Lt. William P., USNR, *Wagner's Men: Stories By and Autobiographies of The Men of Navy Patrol Bombing Squadron (VPB-63)*, Occasioned By The Squadron's Forty-Third Anniversary, Pensacola, Florida, September 19-22, 1985.

Creed, Roscoe, *PBY: The Catalina Flying Boat*, Naval Institute Press, Annapolis, Maryland, 1985

Doenitz, Grand Admiral Karl, *Memoirs: Ten Years and Twenty Days*, Naval Institute Press, Annapolis, Maryland, English Translation 1959, New material 1990.

Gertner, Jon, *The Idea Factory: Bell Labs and the Great Age of American Innovation,* The Penguin Press, New York, New York, 2012.

Hoyt, Edwin P., *U-Boats Offshore: When Hitler Struck America*, Scarborough House/Publishers, Chelsea, Michigan, 1978.

Jones, Richard and Griffin, Patrick, *U-Boats: The Wolf Pack,* Twin Towers Enterprises, Inc., Midwich Entertainment, 1987

Masterman, J.C., *The Double-Cross System in the War of 1939 to 1945*, Yale University Press, New Haven, Connecticutt, 1972.

McCue, Brian, *U-Boats in the Bay of Biscay: An Essay in Operations Analysis*, National Defense University, Washington, DC, 1990

McRaven, William H., *Spec Ops – Case Studies in Special Operations Warfare: Theory and Practice*, A Presidio Press Book, Random House Publishing, 1996.

Meigs, Montgomery C., *Slide Rules and Submarines: American Scientists and Submarine Warfare in World War II*, National Defense University Press, Washington D.C., 1990.

Morison, Samuel Eliot, *History of the United States Naval Operations in world War II: Volume X, The Atlantic Battle Won, May 1943 – May 1945,* Little, *Brown and* Company, Boston, Massachusetts, 1990.

Price, Dr. Alfred, *Aircraft versus Submarines in Two World Wars*, Pen & Sword Aviation, South Yorkshire, England, 2004.

MAD Cats

Scarborough, William, *Walk Around No 5: PBY Catalina,* Squadron/Signal Publications, Carrollton, Texas, 1995.

The Submarine Commander's Handbook, new edition 1943, High Command of the Navy, U.S. Navy translation 1989, Thomas Publications, Gettysburg, PA.

The U-Boat Commander's Handbook, 1943, High Command of the Navy, Thomas Publications, Gettysburg, Pennsylvania, 1989.

Waters, John M., Jr., Captain (ret) USCG, *Bloody Winter*, Naval Institute Press, Annapolis, Maryland, 1967.

Watts, Anthony J., *The U-boat Hunters*, Macdonald and Jane's Publishers, Ltd., London, England, 1976.

www.ingramcontent.com/pod-product-compliance
Lightning Source LLC
Chambersburg PA
CBHW022111150426
43195CB00008B/359